UNDERSTANDING HYDROLOGICAL VARIABILITY FOR IMPROVED WATER MANAGEMENT IN THE SEMI-ARID KARKHEH BASIN, IRAN

UNDERSTANDING HYDROLOGICAL VARIABILITY FOR IMPROVED WATER MANAGEMENT IN THE SEMI-ARID KARKHEH BASIN, IRAN

DISSERTATION

**Submitted in fulfillment of the requirements of
the Board for Doctorates of Delft University of Technology
and of the Academic Board of the UNESCO-IHE
Institute for Water Education
for the Degree of DOCTOR
to be defended in public on
Tuesday, 21 June 2011 at 15:00 hours
in Delft, the Netherlands**

by

Ilyas MASIH

**Master of Philosophy in Water Resources Management,
Centre of Excellence in Water Resources Engineering, U.E.T.,
Lahore, Pakistan
born in Baddomalhi, District Narowal, Pakistan**

This dissertation has been approved by the supervisor:
Prof. dr. S. Uhlenbrook

The research reported in this dissertation has been sponsored by International Water Management Institute, Colombo, Sri Lanka.

CRC Press/Balkema is an imprint of the Taylor & Francis Group, an informa business

Published by:
CRC Press/Balkema
PO Box 447, 2300 AK Leiden, The Netherlands
e-mail: Pub.NL@taylorandfrancis.com
www.crcpress.com – www.taylorandfrancis.co.uk – www.balkema.nl

ISBN 978-0-415-68981-6 (Taylor & Francis Group)

FOREWORD

The right atmosphere required to blaze the way for the desire of obtaining higher education and more and more effectively contributing to the water-related issues was put in my way when I joined the International Water Management Institute (IWMI) as a Junior Hydrologist at IWMI, Lahore, Pakistan in 2001. IWMI's mission to improve the management of water and land resources for food, livelihoods and the environment very much important and captivating and, very soon, I was fully devoted to contribute to the achievement of this sublime mission. During the few years of work at IWMI, I realized how crucial it was to improve the efficiency of water use and raise the productivity of land and water resources to improve food security, ease sectoral competition for water use and safeguard the environment. Furthermore, I found that cross-disciplinary knowledge and understanding constituted an important element of my personal and professional capabilities. At IWMI, where diversity, team spirit and excellence are much appreciated, I learned the importance of striving to attain the best quality of work in the designated area, and tried to understand how small pieces of the puzzle fit together to complete a big picture.

The personal discussion with many colleagues at IWMI, Lahore, Pakistan and Colombo, Sri Lanka, further emphasized the importance of doing PhD studies to better understand and contribute to the abovementioned issues. Dr. Mobin-ud-Din Ahamad remained the cornerstone in this regard, especially because he significantly motivated, guided and recommended me to do PhD studies. The series of discussions with IWMI supervisors and management finally culminated in the form of an offer of a PhD fellowship and to join the IWMI team in Iran on the Karkheh Basin Focal Project (BFP) as a PhD researcher. Dr. Frank Rijsberman, former DG of IWMI and former Professor at UNESCO-IHE Institute for Water Education, Delft, the Netherlands, kindly agreed to be my promoter together with Prof. Stefan Uhlenbrook, Professor of Hydrology at UNESCO-IHE, who also very kindly accepted me as his PhD student.

The BFPs have been very important initiatives of the CGIAR Challenge Program on Water and Food (CPWF), started in several basins worldwide, i.e., Andean, Indo-Gangetic, Karkheh, Limpopo, Mekong, Niger, Nile, Sao Francisco, Volta, and Yellow River, with the main purpose of strengthening the basin focus of the CPWF program. The main aims of the BFPs, including Karkheh BFP, were to provide more comprehensive and integrated understanding of the water, food and environmental issues in a basin; and to understand the extent and nature of poverty within each selected basin and determine where water-related constraints are a major determinant of the poverty factor and where those constraints can be addressed. The adopted research framework was underpinned by the use of sound scientific methods, interdisciplinary knowledge and rigorous research/evaluation methodologies. The scientifically sound knowledge on hydrology and water resources was a substantially important component of the Karkheh BFP beside other areas related to the assessment of water productivity, poverty, institutions and policies. I was designated with the role of conducting a comprehensive assessment of the surface water hydrology that is underpinning the sustainable management of water for food, environment and poverty alleviation.

I strongly consider that the research documented in this thesis has significantly contributed to achieving the project aims and objectives. Moreover, I view that this piece of research is very relevant and beneficial for the hydrological and water management community in Iran and worldwide. This thesis provides an example of understanding issues in local and global contexts, wisely using and further developing existing methods and (scarce) data sets, seeking for innovations to overcome constraints of data, methods and information, and finally realizing the need for having more knowledge and understanding of the variability of hydrological processes and water availability and its proper inclusion in water resources planning and management that envision the well-being of humans and nature.

Ilyas Masih
UNESCO-IHE, Delft, the Netherlands
May 2011

ACKNOWLEDGEMENTS

I give honor and thanks to God, who is the source of knowledge and wisdom, initiator of every good work and leads it to completion, for the provision of necessary intellect and every other resource required for the successful execution of this PhD study.

I am extremely grateful to my promoter Prof. Dr. Stefan Uhlenbrook for his wise guidance, critical and innovative insights, wealth of broad knowledge and understanding, and very strong commitment towards this study that played a pivotal role in the success of this endeavor. I have learned a lot from him professionally as well as personally, which has significantly improved my professional capabilities and greatly enriched my life. His quest for advancing hydrological sciences and their applications, and his dynamic and humble personality, kind and gentle behavior, and ability to stimulate critical thinking and express divergent views in an appealing and inoffensive way are some of the most notable virtues that will remain as exemplary and worth following in the future.

I would like to extend my sincere thanks to my supervisors for all the help they gave me and the guidance they kindly extended to me during this study. I owe many thanks to Dr. Shreedhar Maskey for significantly contributing to this research and admire his consistent encouragement to pursue excellence in every component of this study. I exceedingly benefited from his understanding of hydrological modeling and uncertainty assessment, critical thinking, technical writing ability and computer programming skills.

Many thanks are due to my PhD supervisors from IWMI. Sincere gratitude is due to Dr. Smakhtin for his overall contribution in this research, especially for providing valuable professional insights, guiding in writing skills, kind and generous behavior, and considerate response to all the professional and administrative issues, all of which greatly helped in the successful completion of this study. I would also like to thank my former PhD supervisors from IWMI, Dr. Mobin-ud-Din Ahmad, and Dr. Frank Rijsberman, for making this research directly relevant to IWMI's research portfolio. Both of them had to step out of my PhD supervision, because they left IWMI and joined other organizations. Their supervision and guidance greatly helped achieve a good balance between advancement of hydrological sciences and the practical use of it in the integrated water resources management.

During this study, I worked at IWMI offices in Lahore, Pakistan, Karaj, Iran and Colombo, Sri Lanka, and at UNESCO-IHE, Delft, the Netherlands. I received immeasurably great cooperation from many friends and colleagues at IWMI and UNESCO-IHE. I extend my heartiest thanks to many friends and colleagues of IWMI Pakistan for their support in one way or another, notably to Khalid Mohtadullah, Abdul Hakeem Khan, Zhongping Zhou, Asghar Hussain, Aamir Nazeer, Tabriz Ahmad, Moghis Ahmad, Pervaiz Ramzan, Riaz Wicky and Sidique Akbar. The precious help from a number of people in settling down in Iran, Sri Lanka and the Netherlands, discovering the culture and beauty of these impressive lands is an everlasting treasure for me and my family. I am thankful to Dr. Asad Sarwar Qureshi for his mentoring and extending professional advice and practical support provided during my stay at IWMI Iran and at IWMI Pakistan. Cooperation from Poolad Karimi, in learning Persian and understanding key literature in Persian, data collection, field visits, adjusting in Iran was remarkable and is highly appreciated. I extend my sincere thanks to many colleagues at IWMI Colombo, Sri Lanka who supported me and my family in many ways. Special thanks are due to Dr. Hugh Turral, Dr. Mark Giordano, Dr. Francis Gichuki, Gamage Nilantha, Lal Muthuwatta, Amin-ul- Islam, Mir Matin and Poorna de Silva.

I am thankful for the ever available support and kind interaction with many UNESCO-IHE colleagues. Thanks are due to Susan Graas and Marloes Mul for translating propositions and summary into Dutch. Dr. Ir. Ann van Griensven is acknowledged for her helpful discussions on the SWAT model application. Special thanks to Sylvia van Opdorp-Stijlen, Maria Laura Sorrentino, and Mariëlle van Erven for their cooperation on logistics and counseling issues. I value and appreciate the time spent with many MSc and PhD participants at UNESCO-IHE, and would like to particularly appreciate time spent with Sarfraz Munir, which provided me an excellent opportunity to share my feelings and concerns more openly with someone from my own country, Pakistan. The interaction with many people from various cultures and nationalities at UNESCO-IHE was a unique and highly enriching experience, which has brought added respect for the diversity and difference of opinion and cultures in my life. The friendship with a few Dutch families further helped feel more at home and reduced the agony of missing family and close friends back home. Sincere thanks to Marcel van Genderen and his family, Henk Jansen and his family, Prof. Bill Rosen and many other people from the IREF Church, Delft, for sharing with us in the times of our joys and sorrows and helping us in better understanding the Dutch and European societies.

Funds for this research were generously provided by IWMI through its Capacity-Building Program and from PN 57 'Basin Focal Project for the Karkheh,' a project of the CGIAR Challenge Program on Water and Food, implemented by IWMI in collaboration with several Iranian partners. Additional funds were made available by UNESCO-IHE for supporting me for a few months of additional stay at UNESCO-IHE and for a conference attendance. These funding institutes and their donors are gratefully acknowledged. Special thanks are due to David van Eyck, IWMI Capacity-Building officer, for his gentle and highly professional attitude in executing administrative and financial issues, which greatly facilitated the progress of this study. Similarly cooperation from Ms. Jolanda Boots, PhD Fellowship Officer, at UNESCO-IHE was remarkable and is sincerely acknowledged. Her guidance on social and logistics issues is also highly appreciated.

Main data sets used in this study were accumulated from IWMI data management program, for which kind cooperation of the data management team at IWMI Colombo, Sri Lanka is highly appreciated. Thanks are also due to staff members from IWMI Iran and Sri Lanka offices who exerted every effort in collecting these data sets from the primary and secondary sources. Special gratitude is due to the Ministry of Energy and the Meteorological Organization, Iran, for the provision of necessary data sets on hydrology and climate. Thanks are also due to the Ministry of Jihad-e-Agriculture, in particular to the Agricultural Research and Extension Organization of Iran (AREO) and Soil Conservation and Watershed Management Research Institute (SCWMRI) for their support during data collection, field visits, and providing useful insights on the research issues and results.

I would like to specially thank my parents, wife Huma Ilyas and daughter Sarah Ilyas for their deep love, good care and earnest prayers for me, which provided me with the necessary support, comfort and energy to successfully complete this challenging venture. The sweet company of Huma and Sarah made this tough journey a very pleasant and memorable experience of my life to the extent that I feel very happy to dedicate this work to the two of them.

SUMMARY

The escalating growth of water resources utilization for human purposes, particularly agriculture, is mounting increasing pressure on freshwater resources. Although the human appropriation of water has helped mankind in many ways such as improving food production and socioeconomic well-being, it has also caused damages to the environment and its related services. Balancing water uses for humans and nature is seen as the major challenge of this century. This issue is far more complex for the semi-arid to arid regions of the world, like the Islamic Republic of Iran, where water is generally scarce and demands from agriculture, industry, urbanization and the growing population are rapidly swelling. The high climatic variability and expected ongoing climate change further add to the pressing issues.

Under the condition of water scarcity and competing water uses, improved knowledge of basin-wide hydrology and resource availability are pivotal to instruct informed policy formulation and sustainable development of the water sector. This study is carried out in semi-arid to arid Karkheh Basin of Iran, where massive water allocation planning is on the way, but a comprehensive knowledge on basin hydrology and impact of these developments on different water uses and users across the basin are lacking. The main objective of this research is to provide a hydrology-based assessment of (surface) water resources of the Karkheh Basin and study its continuum of variability and change at different spatio-temporal scales. The methodological framework used in this study was underpinned by the combined use of rigorous system investigation and hydrological modeling techniques. The spatial investigations were carried out at the levels of the river basin, catchment (subbasin) and subcatchment whereas the temporal resolutions were daily, monthly, annually and in long-term time series.

The comprehensive assessment of spatio-temporal variability of surface water hydrology was carried out by using long-term daily streamflow data available for the period 1961 to 2001 for seven important gauging stations located at the Karkheh River and its major tributaries. The analysis was carried out applying techniques, such as measure of central tendency and dispersion, base flow separation and flow duration analysis. Additionally, basin-level water accounting was done for the year 1993-94, for which requisite data sets were available.

The study shows that the hydrology of the Karkheh Basin has high inter- and intra-annual variability, mainly driven by high spatio-temporal variability of climate and spatially diverse soil, land use and hydrogeological characteristics of its drainage area most of which is part of the Zagros mountains. The increase in the streamflows starts in October and lasts till April. Peak flows are normally observed during March-April, but flooding may occur any time between November and April. These high flows are caused by the combined effect of snowmelt and rainfall. The period May through September represent low flows mainly replenished from the base flow contribution from subsurface storages. Moreover, the runoff regime of the middle part of the basin (Kashkan River) is notably different from the upper parts (Gamasiab and Qarasou), with the former showing more runoff per unit area and comparatively higher base flow contributions. The issue of variability is substantiated here by the estimates of mean annual flow and its variability for the Karkheh River gauged at the Paye Pole stations (just downstream of the Karkheh Dam). The mean discharge at this location is 5.83×10^9 m^3/yr., whereas the annual flow was just about one-third (1.916×10^9 m^3/yr.) in the extremely dry year 1999-2000 and as high as 12.60×10^9 m^3/yr. during the highly wet year 1968-69. Under such highly variable conditions, the understanding of the reliability of the water availability becomes more meaningful for better resources use and allocation decisions. The flow duration analysis conducted in this study provides such estimates of streamflow reliability for the Karkheh Basin at daily, monthly and annual time resolutions.

The synthesis of the results on hydrological variability, water availability, and water accounting suggests that the Karkheh Basin was an open basin during the study period (1961-2001), and there is further room for water resources allocations, i.e., in the range of 1-4 $\times 10^9$ m^3/yr. depending on the amount of water left for environmental flows. However, the allocation should be done after a careful impact assessment and trade-off analysis for multiple and highly competing uses and users across the basin. The evaluation of ongoing water allocation planning appeared as nonsustainable given the limitations of resources availability and its high variability. If the current water policy is implemented the basin will soon approach the closure stage in the near future (latest by 2025), and then, meeting demands of all users will be extremely difficult, especially during low flow months and dry years. The environmental sector is likely to suffer the most which, so far, has been given low priority, but other sectors such as agriculture and domestic uses are also likely to face reductions in their allocated water rights.

The changes in the hydro-climatic variables and their linkages were also explored as part of the system analysis. Streamflow records from five mainstream stations were used for the period 1961-2001 to examine trends in a number of streamflow variables representing a range of flow variability, i.e., mean annual and monthly flows, 1 and 7 days maximum and minimum flows, timing of the 1-day maxima and minima, and the number and duration of high- and low-flow pulses. Similarly, the precipitation and temperature data from six synoptic climate stations were used for the period 1950s to 2003 to examine trends in climatic variables and their correlation with the streamflow. The Spearman rank test was used for the detection of trends, and the correlation analysis was based on the Pearson method. The results revealed a number of significant trends in streamflow variables both increasing and decreasing. Moreover, the observed trends were not spatially uniform. The decline in low-flow characteristics were more significant in the upper parts of the basin (particularly for Qarasou River), whereas increasing trends in high flows and winter flows were noteworthy in the middle parts of the basin (Kashkan River). Most of these trends were mainly attributed to precipitation changes. The results showed that the decline in April and May precipitation caused decline in the low flows while increase in winter (particularly March) precipitation coupled with temperature changes led to an increase in the flood regime. The observed trends at the Jelogir station on the Karkheh River reflect the combined effect of the upstream catchments. The significant trends observed for the number of streamflow variables at Jelogir, e.g., 1-day maximum, December flow and low pulse count and duration, indicated alterations of the hydrological regime of the Karkheh River and were mainly attributed to the changes in the climatic variables.

Regionalization of hydrological parameters emerged as an important issue for the Karkheh Basin because streamflow records were not available for many subcatchments, and many streamflow gauging stations were abandoned. A new regionalization method was developed in this study to estimate streamflow time series for poorly gauged catchments. The proposed method is based on the regionalization of a conceptual rainfall-runoff model based on the similarity of flow duration curves (FDC). The performance of this method was compared with three other methods based on drainage area, spatial proximity and catchment characteristics. The data of 11 gauged catchments (475 to 2,522 km^2) were used to develop the regionalization procedures. The widely used HBV model was applied to simulate daily streamflow with parameters transferred from gauged catchment counterparts. The study indicated that transferring HBV model parameters based on the FDC similarity criterion produced better runoff simulation compared to the other three methods. Furthermore, it was demonstrated that the parameter uncertainty of the model has little impact on the regionalization outcome. The results of this novel method compared very well with most of the promising regionalization techniques developed and applied elsewhere. Therefore, the FDC-based model regionalization method developed in this study is a valuable addition to existing regionalization methods. The proposed method is easy to replicate in other river basins, particularly those facing a declining streamflow network.

Furthermore, a semi-distributed, process-based model – Soil Water Assessment Tool (SWAT) – was used to understand and quantify the hydrological fluxes, and to test different scenarios. It was recognized that the widely used SWAT model offers a range of possibilities for defining the model structure, but the input of climatic data is still rather simplistic. SWAT uses the data of a precipitation gauge nearest to the centroid of each subcatchment as an input for that subcatchment. This may not represent overall catchment precipitation conditions well, and may lead to increased uncertainty in the modeling results. In this study, an alternative method for precipitation input was evaluated. In particular, the input of interpolated areal precipitation was tested against the standard SWAT precipitation input procedure. The extent of the modeling domain was 42,620 km^2, located in the mountainous semi-arid part of the study basin, from where almost all of the basin's runoff is generated. The model performance was evaluated at daily, monthly and annual scales using a number of performance indicators at 15 streamflow gauging stations, each draining an area in the range of 590-42,620 km^2. The comparison suggested that the use of areal precipitation improves model performance particularly in small subcatchments with drainage area in the range of 600-1,600 km^2. The areal precipitation input results in increased reliability of simulated streamflows in the areas of low rain gauge density and poor spatial distribution of the rain gauge(s).

Both precipitation input methods result in reasonably good simulations for larger catchments (over 5,000 km^2), which was attributed to the averaging out effect of precipitation at larger spatial resolution.

The understanding of catchment hydrology through the abovementioned studies, field visits and literature review, and rigorous parameter estimation procedures helped achieve reasonably good calibration, validation and uncertainty analysis of the SWAT model for the Karkheh River Basin. This provided adequate confidence for using the SWAT model for the analysis of water use scenarios in the basin. Three scenarios, related to increased water use in rain-fed agriculture, were evaluated. The tested scenarios are: upgrading rain-fed areas to irrigated agriculture (S1), improving soil water availability through rainwater harvesting (S2), and a combination of S1 and S2 (S3). The results of these scenarios were compared against the baseline case over the study period 1988-2000. The baseline simulations were carried out using the finally adopted model structure and a parameter set obtained from the used calibration procedure. The results of the first scenario (S1) indicated a reduction of 10% in the mean annual flows at the basin level, which ranged from 8 to 15% across the main catchments across the basin. The reductions in the mean monthly flows were in the range of 3-56% at the basin level. The months of May-July sustained high impacts, with June witnessing the highest percentage of flow reductions. Flow reductions in these months were more alarming in the upper parts of the basin which was mainly attributed to relatively higher potential of developing rain-fed area to the irrigation, coupled with comparatively lower amounts of runoff available in these months. The impacts of S2 were generally small at the catchment as well as basin scale, with reductions in the range of 2-5% and 1-10% in the mean annual and mean monthly flows, respectively. The estimated flow reductions at the annual scale remain well within the available water resources development potential in the basin. However, avoiding excessive flow reductions in May-July would require adoption of additional measures, such as practicing supplementary irrigation and augmenting supplies through developing a range of water storage options, and considering less than the potential rain-fed area for upgrading to irrigated farmland (particularly in upper parts of the basin).

The study concludes that understanding of the prevalent high level of variability in hydrology and water resources, a sound foundation of which has been laid by this study, and inclusion of a range of variability of the water resources into planning and management does play a pivotal role in the sustainable use and management of available water resources of the Karkheh Basin. The ongoing water allocation planning is not sustainable and a thorough revision of it is recommended, which will essentially require the reduction in water allocations to human uses (particularly agriculture) and leaving more water for the environment. The climate variability and change have significantly altered the hydrological regime of the Karkheh River system, warranting immediate mitigation efforts, i.e., structural measures and programs to reverse catchment degradation to manage intensified flood regime in the middle parts of the basin and considering how to reduce water withdrawals during low-flow months (May to September) in upper parts of the basin in order to mitigate the impacts of declining low flows in these areas. The impact evaluation study conducted herein have shown that the improving water use in rain-fed agriculture could be promoted in the basin, with consideration of in-situ soil and water conservation interventions all across the basin as they pose minimal impacts on downstream water availability. However, the conversion of rain-fed areas to irrigation requires a cautious approach to ensure reasonable levels of flow reduction on monthly time resolution, which calls for upgrading limited rain-fed areas to irrigation (particularly in upper parts of the basin), practicing supplementary irrigation and developing a range of water storage options. Strengthening hydro-climatic data- monitoring networks is recommended to improve available data and consequent application of hydrological and water management models for more informed decision-making processes. In this regard, rehabilitation of abandoned hydro-climatic stations and consideration of installation of more monitoring stations in the mountainous parts are recommended. Planning and managing all water resources in the river basin context should be promoted in the study basin.

In general, the knowledge generated in this case study is very much relevant for other river basins of Iran, and worldwide.

TABLE OF CONTENTS

1.   INTRODUCTION

1.1.   Background

1.1.1.   Increasing pressure on earth's water resources

Water plays a key role in sustaining life on our planet earth. We use water not only for our basic survival (e.g., for drinking, cooking, bathing and sanitation) but also for many other purposes such as hydropower generation, industry, navigation and recreation. Water is essential not only for meeting human needs but for nature where it is essentially required to maintain fisheries, wildlife, riparian vegetation, river deltas and aquatic biodiversity.

The freshwater resources of the earth are finite and are distributed into hydrological storages as glaciers, groundwater, freshwater lakes and wetlands, soil moisture, atmospheric water and river waters (Shiklomanov and Rodda 2003). Balonishnikova et al. (2006) have estimated that the total renewable freshwater resources of the world are about 42,700 km^3/yr.. The spatio-temporal distribution of water is very much nonuniform across the globe. Also the full amount of renewable water is not accessible to human uses due to different reasons such as the fact that a major part of the rainwater flows as flood runoff during short period of time. This high spatio-temporal variability together with extreme climatic events in the form of floods and droughts, and localized high demands from intensive agriculture and big cities make water management a very complicated task.

Large investments in infrastructure (e.g., dams and irrigation facilities) have resulted in a rapid increase in the uses of water for human purposes during the last century (Figure 1) (Shiklomanov 1999). The major share of the total water withdrawals and consumptions pertain to the agriculture sector (about 70%) followed by industrial and municipal sectors. The world water withdrawals have increased over 7 times during the last century, i.e., from 578 km^3/yr. in the year 1900 to about 3,788 km^3/yr. in the year 1995. This trend is projected to continue in future, though with comparatively lower rates. As a consequence, the freshwater resources of the world are under ever-increasing pressure due to escalating demands. The main driving forces behind this rising pressure are: population growth; major demographic changes as people move from rural to urban environments; higher demands for food security and socioeconomic well-being; increased competition between uses and usages; and pollution from industrial, municipal and agricultural sources, climate variability and change, and land use change (e.g., WWAP 2006).

Despite the immense progress in water development the demands are still very difficult to meet in many regions of the world. There are about 1.1 billion people who still do not have access to improved water supply, and about 2.4 billion, i.e., 40% of the world population, lack access to improved sanitation (WHO 2000).

Irrigated agriculture has to expand further to meet the food needs of growing populations and hence withdrawals to irrigated agriculture will keep increasing (Seckler et al. 1998). Even if irrigation efficiency could be improved dramatically at some places, meeting these water demands will be a big challenge in many parts of the world, especially in developing countries of Asia and Africa as these regions will be facing severe water scarcity in the coming decades (Rijsberman 2006).

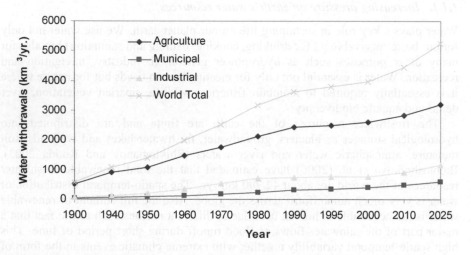

Figure 1. Trends in the global water withdrawals by sector of economic activity.
(Data source: Shiklomanov, 1999, cited in Cosgrove and Rijsberman 2000a.)

A historical overview depicts that the human appropriation of freshwater water resources has helped in many ways such as preventing food crises in the world, provision of water and sanitation, generating electric power and mitigation of damage from hydrological hazards such as flood and drought. But it is now well recognized that water resources strategies of the last century have largely worked against nature and have resulted in environmental degradation as many rivers no longer reach the sea for extended periods of time, river delta regions are ruining, groundwater in the world's key aquifers are depleting, water pollution is increasing and aquatic ecosystems are being increasingly damaged (Rijsberman and Molden 2001; Gleick 2003; Postel and Richter 2003). Many countries of the world are facing this conflicting situation at present and are searching for sustainable solutions to achieve a balance among human and ecosystem uses of water. However, most of the restoration examples are limited to USA, Australia, South Africa and Europe (Tharme 2003; Smakhtin et al. 2004). Balancing water for human needs and for nature is a big challenge faced by many countries at present and has been regarded as one of the greatest challenges of this century all across the globe (Rijsberman and

Molden 2001; Zehnder et al. 2003; Postel 2003; Loucks 2006; Palmer and Bernhardt 2006).

1.1.2. Adapting sustainable solutions

Water issues in the world are diverse in nature, governed by a large array of natural and anthropogenic forces such as climatic conditions, land features, hydrological behavior, variability of water resources, socio-political conditions, economic factors, technological capacity and ecosystems needs. Integrated Water Resources Management (IWRM) has been advocated as the better way forward for addressing the complex and dynamic nature of the water-related issues (e.g., Bouwer 2000; Karamouz et al. 2001; Snellen and Schrevel 2004; van der Zaag 2005; Savenije and van der Zaag 2008).

Emphasizing the adoption of an integrated approach to water resources management, Cosgrove and Rijsberman (2000a) suggested that limiting the expansion of irrigated agriculture, increasing water productivity, developing biotechnology for agriculture, increasing storage, reforming water resource management institutions, increasing cooperation in international basins, valuing ecosystem functions and supporting innovation would be the key areas of interventions contributing towards addressing the global water crisis and, consequently, would help achieving the Millennium Development Goals (MDGs).

Gleick (2003) argued that the "hard path[1]" solutions of the past are no longer better choices and we need to follow the "soft path[2]" solutions. Therefore, we need to rely on carefully planned and managed centralized infrastructure complimented by small-scale decentralized facilities; strive for improving the productivity of water rather than seeking for endless sources of new supply; deliver water services and qualities matched to users' needs rather than just delivering quantities of water; apply economic tools for promoting efficient water use; and include local communities in decisions about water management, allocation and use.

Vörösmarty et al. (2000) have recommended that an integrated research on climate change, water resources and socioeconomic aspects would be essential for making progress as the population growth and economic development will be the main forces escalating the water demands in the future. Investments in socioeconomic and hydrometric data are important and should be enhanced for making adequate progress.

Improving productivity of water in agriculture is regarded as one of the most promising solutions (CAWMA 2007). It is argued that producing more food with

---

[1] Hard path refers to the approach based on the construction of massive infrastructure in the form of dams, aqueducts, pipelines, and complex centralized treatment plants, which dominated the water agenda of the twentieth century.

[2] Soft path refers to the approach based on the carefully selected centralized physical infrastructure with lower-cost community-scale systems, decentralized and open decision making, water markets and equitable pricing, application of efficient technology, and environmental protection (Gleick 2003).

less or with the same amount of water (*more crop per drop*) will lead to more food security, less infrastructural requirements, reduced competition for water as less water will be needed for agriculture and more can be diverted for domestic, industrial and environmental purposes (Cosgrove and Rijsberman 2000a, b; Postel 2000; Rijsberman and Molden 2001; CPWF 2002). Improving productivity of water both from rain-fed and irrigated lands is a key focus of the new blue-green water paradigm (Falkenmark and Rockström 2006).

The list of potential solutions is quite long. Just to mention a few more: creating awareness among all the stakeholders about water-food-environment nexus (DIALOGUE 2002) and developing and adopting new technologies and changing lifestyles (e.g., changing dietary patterns, improving education and reducing population growth rates) would be very essential for matching the water supplies and demands in the future (Gallopin and Rijsberman 2000; Cosgrove and Rijsberman 2000b). There is a need to change mindsets, policies and practices and to overhaul water policies and practices in a way that will protect freshwater ecosystems and their valuable services (Postel 2003, 2005). While we update water policies, the highest priority should be given to the following three policy areas (Postel 2005): a) securing drinking water supplies through increased investments in the catchment protection; b) inventorying and setting ecological goals for the health of rivers, lakes, and other freshwater ecosystems and establishing caps on the degree to which human activities are allowed to modify river flows, deplete groundwater, and degrade catchments; and c) improving water productivity both from agriculture and nature through a combination of efficient water use and implementation of caps on water use.

1.1.3. *Managing water by river basin*

Scale consideration is very important both for the understanding and simulation of hydrology (Blöschl and Sivapalan 1995) and for the management of water resources (Zehnder et al. 2003; van der Zaag and Gupta 2008). Water issues and water management could be viewed in many different spatial scales such as global, continent, country, river basin, catchment (subbasin), irrigation system, city, wetland, farmer field, etc. The temporal scales could be every minute, hour, daily, month, season, decade, year or even every specified longer period. It is now well recognized that the river basin is the most appropriate scale for the sustainable management of water resources (WWAP 2006; Molle 2006). The European Frame Work directive is a well known example in this regard which states that the rivers, lakes and groundwater resources need to be managed by the river basin which is a natural hydrological unit, instead of only according to administrative or political boundaries (Ringeltaube 2002).

It must be noted that there are a lot of unknown processes and facts pertaining to all of the abovementioned scales. It is extremely important to understand the present state of a river basin with respect to the degree of "basin closure,"[3] i.e., whether it is an open basin, closed basin or closing basin, as this knowledge has implications for many water management polices (Keller et al. 1996, 1998; Seckler, 1996; Falkenmark and Molden 2008). For instance, adapting water conservation and irrigation efficiency improvement strategies aiming at water savings may not really save water in a closed basin and may merely reflect the reappropriation among different users/uses. In such cases, improving overall productivity of water is a more plausible alternative. Furthermore, understanding various factors, such as hydrological, water management, socio-political and economic, governing the river basin transformations and water uses are also essential (Molle 2003).

Management of water resources from a river-basin perspective requires comprehensive interdisciplinary analysis, evaluation of present and future conditions, and formulation of multiple management plans (Schultze 2001). But, there are several scientific and technical obstacles that prohibit us from understanding, predicting and ultimately guiding the management of water resources. The major scientific issues are the lack of understanding of hydrological processes at the basin scale and inadequate understanding of the coupling between hydrological, ecological and climate systems (Uhlenbrook 2006; Uhlenbrook et al. 2006).

1.1.4. Need for a hydrological synthesis

The need for hydrological investigations was at the core of the hard path solutions of the last century. The hydrology-based assessment of water resources was then integrated with the information from other disciplines (such as geology, soil science, atmospheric science, sociology/anthropology and various engineering disciplines) for implementing water resources development and management strategies. The pivotal role of hydrology in implementing the hard path solutions is quite evident and has been very well internalized in the planning, construction and operational phases of the water resources development projects. However, in the past when structural measure were the main options for solving water availability issues, this role of hydrology was much simpler, as water was abundant and the effects of

---

[3] A water resource system is "closed" when there is no usable water leaving the system other than that necessary to meet minimum instream and outflow requirements (Keller et al. 1998). From the agricultural standpoint, either all of the initial available water supply has been lost to beneficial evaporation and crop evapotranspiration (ET_c), plus unavoidable nonbeneficial evaporation and ET_c, or it has such high concentrations of salts and other pollutants that it is unusable. Conversely, an integrated water resource system is "open" when excess usable water does leave the system and there is nonbeneficial evaporation and ET that can be avoided. According to a recent definition by Falkenmark and Molden (2008), a river basin is termed closed when additional water commitments for domestic, industrial, agricultural, or environmental uses cannot be met during all or part of a year, while in an open basin more water can be allocated and diverted.

anthropogenic causes, climate and land use changes were not that prominent. But the scientific role of hydrology is much more demanding and challenging now and in the future when water challenges are more pronounced, diverse and complicated.

Hydrological investigations are essentially needed before formulating both hard and soft path solutions and should be continuously updated in view of changing needs and conditions. Understanding of hydrological processes and hydrology-based assessment of water resources and water balance is, in fact, an integral part or a basic requirement for most of the abovementioned solutions. For instance, it is one of the essential components of water productivity estimations (Molden 1997) and water scarcity studies (Seckler et al. 1998; Sullivan et al. 2000). Hydrological investigations are essentially required to devise action plans for the policy areas as proposed by Postel (2005) on sustainable uses of water by humans and ecosystems (see section 2.1). For example, putting caps on water withdrawals requires a quantitative assessment of the availability of water resources. We need to study/model the hydrological response of the catchments before making catchment restoration measures and investment decisions and we essentially require detailed information on the spatio-temporal pattern of water flows for the restoration of the natural hydrological regime of the rivers.

Similarly, hydrological analyses at basin and smaller levels are required to obtain the knowledge of the degree of a basin closure and flow paths of water which can then help in guiding the kind of appropriate interventions. Without such hydrological assessments much of the debate on the real water savings, upstream- downstream impacts (Keller et al. 1996 and 1998; Seckler 1996; Molle et al. 2004) or whether a demand-side or supply-side intervention is better (Molle and Turral 2004) remains mere conceptual and, therefore, a qualitative assessment of complex realities may lead to erroneous planning and consequently nonsustainable management of water resources.

The issues of water resources management are becoming increasingly important almost everywhere in the world and water-related problems are becoming increasingly complex. The role of hydrological investigations remains pivotal in exploring sustainable solutions for the present and emerging water issues. A hydrologic synthesis in at least three respects is essentially required, which are across a) processes, where the challenge is how to represent complex interacting dynamic systems including feedback between system components, b) places, where the challenge is how to synthesize the plethora of case studies around the world in the past decades, and c) scales, where one is interested in the general characteristics of processes as a function of spatio-temporal scales for the same site or an ensemble of sites (Blöschl 2006).

Therefore, there is need to increase understanding of hydrological processes in spatio-temporal scales and their interaction with humans and ecosystems. Then the challenging undertaking, for future research and practices in hydrology and water resources, is how to induce and deduce sustainable water management strategies based on the hydrological research.

1.2. Hydrological and Water Management Issues in the Karkheh Basin, Iran

1.2.1. An overview of the water issues of Iran

The Islamic Republic of Iran is located in Southwest Asia and is situated between approximately 25-40 degrees northern latitudes and 44-64 degrees eastern longitudes. The total area of Iran is about 1.65 million km^2, out of which about 52% is mountainous and desert terrain and about 16% is terrain with an elevation of over 2,000 meters above sea level (masl) (FAO 1997). The two largest and highest mountain systems are Zagros and the northern highlands (Talish and Alburz), the former extending from northwest to southeast, while the later stretches from west to east along the southern Caspian Sea.

Forests and woodlands comprises only 7% (11.4 million ha) of the total land area and about 27% (44 million ha) is under pastures and meadows (FAO 2006). The arable land and permanent crops are estimated to be 16.1 and 2.1 million ha, respectively. The agricultural area under irrigation has grown from 4.7 million ha in 1961 to 7.7 million ha in 2003, indicating a growth of about 63% over this period. Despite tremendous increase in irrigated area, rain-fed farming is a very important feature of the country's food security and agricultural economy.

The climate of Iran depicts extreme variations due to its geographic locations and varied topography. Generally, it is regarded as a country of dry conditions and its climate is mostly arid to semi-arid. Precipitation (P) patterns show large spatial and temporal variations, caused mainly by Zagros and the northern mountain ranges. The average annual P over Iran is about 240 mm/year. (/yr.) (Dinpashoh et al. 2004). Over half of the country's area receives less than 200 mm/yr. and over 75% receives less than 300 mm/yr.. Only 8% of the area receives more than 500 mm/yr.. The seasonal distribution in winter (January-March), spring (April-June), summer (July-September) and autumn (October-December) is about 53, 20, 4 and 23%, respectively, of the annual P.

Iran has several large rivers, among which Kurun, Dez and Karkheh are the three major ones. Most of the rivers and streams are steep and irregular and end up in the marshes/wetlands. Most of the marshes and wetlands of Iran have high significance for their biodiversity, environmental and socioeconomic values. Water is also stored naturally underground both in confined and unconfined aquifers, finding its outlet in *qanats* (subterranean water canals), springs and streams. Vakili et al. (1995) analyzed the different estimates of water resources of Iran and suggested that the total quantity of renewable water resources is about 135 km^3/yr.. According to FAO (1997), the internal renewable water resources of Iran are estimated at 128.5 km³/yr.. It receives 6.7 km³/yr. of surface water from Helmand River having a drainage area in Pakistan and Afghanistan. The flow of the Arax River, at the border with Azerbaijan, is estimated at 4.63 km³/yr.. Surface runoff represents a total of 97.3 km³/yr. whereas groundwater recharge is estimated at around 49.3 km³/yr. of which 12.7 km³/yr. are obtained from infiltration from the riverbeds.

Consistent with the global trends shown in Figure 1, the increasing water withdrawals continue to amplify pressure on the water resources of Iran (Figure 2).

Water withdrawals have doubled over the last 3 decades, rising from 45 km^3/yr. in 1975 to 93.3 km^3/yr. in 2004 (FAO 2009). Most of this increase is due the increased allocations to the agriculture sector.

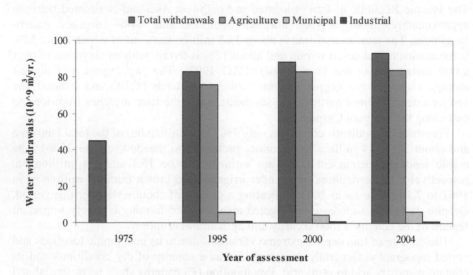

Figure 2. Water withdrawals by sector in Iran.
(Data source: FAO 2009, Aquastat database.)

The major driver of this trend is the country's policy to attain food self-sufficiency illustrated in Figure 3, showing the increase in cereal area and cereal production in the country. However, despite the increasing trend in the production over time, the cereal import was imperative to meet demands (FAO 2009). The high variations in cereal area, yield and imports could be attributed to the variable nature of the climate and water resources. For instance, the food production faced serious decline in the dry years, 1999-2001, and therefore about 10 million tonnes of cereals were imported costing about 1.5 billion US dollars in 2000.

Various sources project that Iran would be facing serious water stress and water scarcity problems by the first quarter of this century (Seckler et al. 1998; Wallace 2000; Alcamo et al. 2000; Sullivan et al. 2000; Yang et al. 2003). Figure 4, showing trend in population increase and corresponding decline in per capita water availability, demonstrates the simplest representation of water scarcity. The water availability/capita/yr. was about 6,057 m^3/person/year in 1961, which showed a sharp decline of about 70% over the period 1961 to 2009, reaching about 1,820 m^3/year/person in 2009. Given the rising trends in population, the per capita water availability is projected to fall below the water stress threshold value of 1,700 m^3/person/year in the coming decade by as early as 2015.

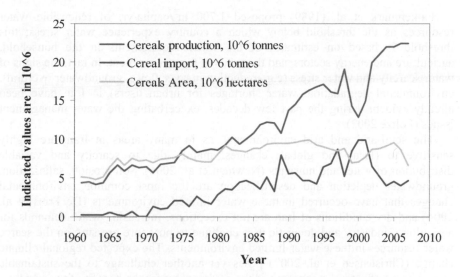

Figure 3. Cereal area, production and import in Iran during 1961-2007.
(Data source: FAO 2009.)

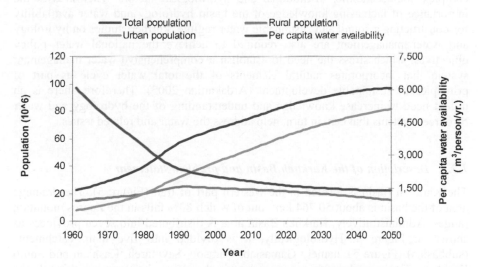

Figure 4. Overview of trends in per capita availability of renewable water resources and population growth of Iran (1961-2050).
(Data for population estimate and projections are taken from FAO 2009 whereas the value of 135 km³/yr. is used as annual renewable water resources after Vakili et al. 1995.)

Falkenmark et al. (1989) proposed 1,700 m^3/capita/yr. of renewable water resources as the threshold below which a country experience water stress; this threshold is based on estimates of the water requirements in the household, agriculture and energy sectors, and the needs of the environment. In fact, the signs of water scarcity and water stress (e.g., reduction in river flows, groundwater overdraft, environmental degradation, water shortages for urban users) in Iran have been already evident during the past few decades, exacerbating the water management issues (Foltze 2002).

The semi-arid and arid environments, as in many areas in Iran, are highly sensitive to (local and global) changes, mainly due to scarcity and variable distribution of water and nutrients (Newman et al. 2006). Soil erosion, salinization, groundwater depletion and desertification are the most common environmental changes that have occurred in these water limited environments (De Fries et al. 2004) and dry conditions of Iran are not exceptions. Increasing water demands for agriculture, industry and domestic uses continue to put more pressure on the scarce water resources in these water-limited environments. The expected regional climate change (Christensen et al. 2007) poses yet another challenge to the sustainable management of natural resources and the environment for the benefit of the society.

In summary, the water crisis of Iran is likely to intensify given the increasing competition of water for human uses and the environment. There are many other governing factors ranging from natural and anthropogenic climate changes to the complex socioeconomic, institutional and hydrological factors. This stresses the importance of increasing knowledge of the basin hydrology and water availability for constructing a sound and sustainable water regime. Further studies on hydrology and water management are also required to achieve the national water policy objectives, which stress the need to establish a comprehensive water management system that incorporates natural elements of the total water cycle as part of principles of sustainable development (Ardakanian 2005). Therefore, there is an urgent need to increase knowledge and understanding of the hydrology and water resources systems that can, in turn, help address the water and related issues.

1.2.2. Description of the Karkheh Basin and problem statement

The Karkheh Basin is located in the western part of Iran (Figure 5). The drainage area of the basin is about 50,764 km^2, out of which 80% falls in the Zagros mountain ranges. Administratively, Karkheh Basin area is distributed into seven provinces as shown in Figure 5. Hydrologically, it is divided into five main catchments (subbasins) (Figure 5), namely Gamasiab, Qarasou, Saymareh, Kashkan and South Karkheh. These catchments are named after the main river passing through the respective areas. The Karkheh River eventually terminates in the Hoor-Al-Azim swamp, a large transboundary wet land shared with Iraq, which is connected to Euphrates-Tigris system.

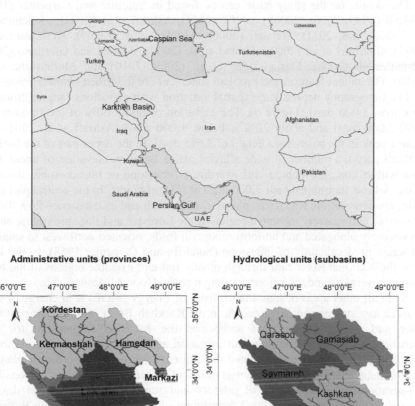

Figure 5. Location of Karkheh Basin in Iran and its hydrological and administration units.

The details on the study basin can be found in Sutcliffe and Carpenter (1968), JAMAB (1999; 2006a, 2006b), UNEP (2001), Ashrafi et al. (2004), , Karamouz et al. (2006; 2008, 2011), Heydari (2006), Absalan et al. (2007), Keshavarz et al. (2007), Ghafouri et al. (2007), Ahmad et al. (2009), Ahmad and Giordano (2010), Marjanizadeh (2008), Marjanizadeh et al. (2009; 2010) and Muthuwatta et al. (2010). The salient features and problem statement are described below.

The topography depicts large spatial variation with elevations ranging from 3 to more than 3,000 masl (Figure 6). The elevation of about 60% of the basin area is 1,000-2,000 masl and about 20% is below 1,000 masl (Ashrafi et al. 2004). The highest peak in the basin has a height of 3,645 masl. In the upper part of the basin (in northern parts), a number of wide alluvial plains lie at an elevation of about 1,500 masl within complex faulted and overthrust limestone or metamorphic mountain masses whose summit exceeds 3,000 masl at several points. In the central part of the basin, upstream of the Khuzestan plains, the Karkheh and its tributaries flow through the remote and sparsely inhabited region of the Lorestan and Ilam provinces, an area of extremely elongated and uniform mountain folds, oriented northwest to southeast and again predominantly of limestone (Sutcliffe and Carpenter 1968). In the lower parts, the Karkheh River runs through mostly flat and irrigable regions of the basin, through several meanders, before draining into the Hoor-Al-Azim Swamp.

As in all other areas of Iran, the Ministry of Energy (MOE) is in charge of water resource assessment and development in the Karkheh Basin. Through its provincial water and power development authorities the MOE is responsible for large hydraulics works, including dam and irrigation and drainage canals for distribution of water. MOE and its water-related department oversee and coordinate planning, development, management and conservation of water resources. The responsibility of operation and maintenance of primary and secondary irrigation and drainage canals lies within the water-related department of MOE. The Khuzestan Water and Power Development Authority (KWPA) is among the key institutions dealing with water issues in the Karkheh Basin. The Ministry of Jihad-e-Agriculture, through its provincial organizations, is responsible for on-farm water management, on-farm irrigation and drainage networks, rain-fed and irrigated crops, catchment management and other related issues. Many other social and nonformal institutions are functioning in the basin; working for the local water management activities. These local organizations have derived their water allocation and management principles through the rich history of Iranian cultures.

The population living in the basin is about 4 million (in 2002), and about one-third resides in the rural areas (JAMAB 1999; Ashrafi et al. 2004). The annual population growth rate is about 2.6%. Historically, the Karkheh Basin had been the cradle of ancient civilization of Mesopotamia and a boundary between Arab and Persian cultures. The Karkheh Basin, once called the "breadbasket of Southwest Asia" now faces many challenges such as low water and land productivity, poverty, land degradation, groundwater depletion and growing competition for water among upstream and downstream areas and among different sectors of water use such as irrigation, domestic, hydropower and environment (CPWF 2003).

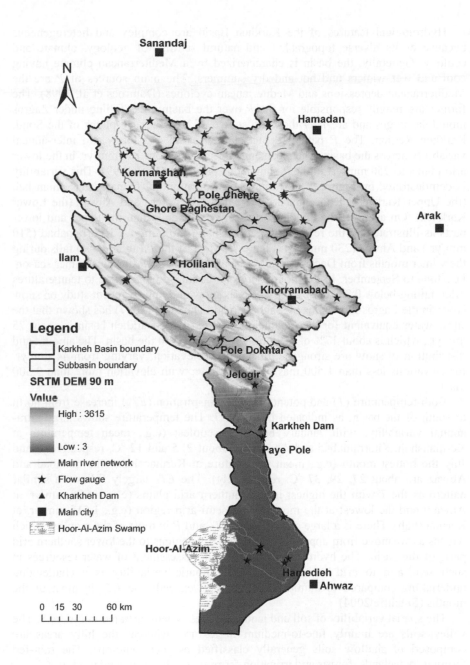

Figure 6. Digital elevation map of the Karkheh Basin and the streamflow monitoring network.

Hydrological features of the Karkheh Basin are complex and heterogeneous because of its diverse topography, and natural settings of geology, climate and ecology. Generally, the basin is characterized by a Mediterranean climate having cool and wet winters and hot and dry summers. The main sources of P are the Mediterranean depressions and Mediterranean cyclones (Domroes et al. 1998). The former are mainly responsible for the P over the basin areas falling under Zagros mountain ranges and are later the main source of P in the arid plains of the South Karkheh Region. The P pattern depicts large spatial and intra- and inter-annual variability across the basin. The mean annual P ranges from 150 mm/yr. in the lower arid plains to 750 mm/yr. in the mountainous parts (JAMAB 1999). This variability is demonstrated by Figure 7 indicating the mean monthly climate of Kermanshah (the Upper Karkheh), Khorramabad (the Middle Karkheh) and Ahwaz (the Lower Karkheh). On average, the middle part receives higher P than the upper and lower parts as illustrated by the records of Kermanshah (450 mm/yr.), Khorramabad (510 mm/yr.) and Ahwaz (230 mm/yr.) (Figure 7). Most of the P (about 65%) falls during the winter months from December to March and almost no P during summer season, i.e., June to September. In the mountainous parts during winter, due to temperatures often falling below 0 °C, the winter P falls as snow and rain. A recent study on snow cover in the Zagros mountains by Saghafian and Davtalab (2007) has shown that the snow water equivalent for the mountainous parts of the Karkheh basin is about 75 mm/yr., which is about 17% of the long-term annual P in the basin. The amount and distribution of snow are strongly influenced by elevation, varying from 44 mm/yr. for elevations less than 1,500 masl to 245 mm/yr. with elevation more than 3,500 masl.

Both temperature (T) and potential evapotranspiration (ET_P) increase from north to south of the basin, as indicated in Figure 7. The temperature shows large intra-annual variability, with January being the coolest (e.g., mean temperature at Kermanshah, Khorramabad and Ahwaz are about 2, 5 and 12 °C, respectively) and July the hottest month (e.g., mean temperature at Kermanshah, Khorramabad and Ahwaz are about 27, 29, 37 °C, respectively). The ET_P largely follows a similar pattern as the T with the highest in the southern arid plains (e.g., 1,930 mm/yr. at Ahwaz) and the lowest at the mountainous semi-arid region (e.g., 1,515 mm/yr. at Kermanshah). There is a large gap between ET_P and P in most of the months, which widens as we move from upper northern semi-arid regions to the lower southern arid parts of the basin. The hydrological analysis and assessment of water resources in such semi-arid to arid regions with high climatic variability is a challenging undertaking compared to humid areas where P exceeds the ET_P in most of the months (Sutcliffe 2004).

The spatial variability of soil and land use types is demonstrated in Figure 8. The valley soils are mainly fine-to-medium in texture, whereas the hilly areas are composed of shallow soils generally classified as rock outcrops. The rain-fed farming, rangelands, forests and irrigation farming are the main land use types.

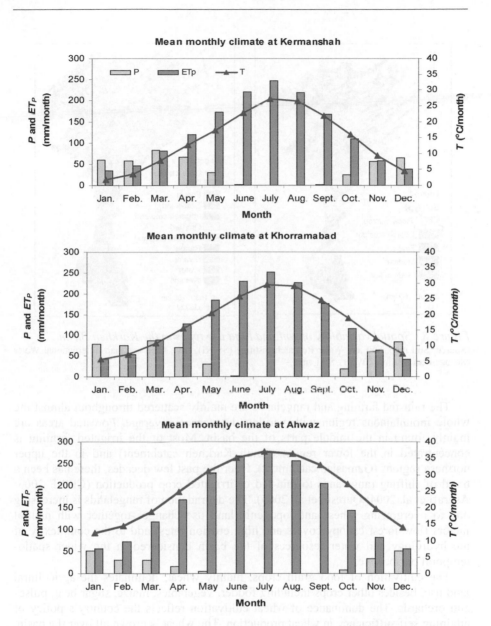

Figure 7. Mean monthly climate of the Karkheh Basin, illustrated by precipitation (P), temperature (T) and potential evapotranspiration (ETₚ) at the three climatic stations Kermanshah, Khorramabad and Ahwaz. Data source: Meteorological Organization of Iran.

(The averages are for the period of 1950s to 2003. potential evapotranspiration was estimated using Hargreaves method, Hargreaves et al. 1985).

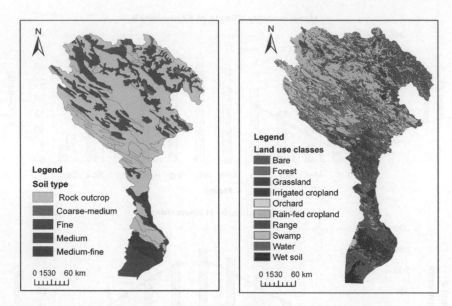

Figure 8. Spatial variability of soil and land use types in the Karkheh Basin.
(Source: Soil map: Soil and Water Research Institute (SWRI), Iran; Land use map: International Water Management Institute (IWMI), Sri Lanka.)

The rain-fed farming and rangelands are mainly scattered throughout almost the whole mountainous region with varying degrees of coverage. Forested areas are mainly found in the middle parts of the basin. Most of the irrigated farming is concentrated in the lower region (South Karkheh catchment) and in the upper northern regions (Gamasiab catchment). Over the past few decades, there has been a trend of shifting rangelands to rain-fed or irrigated crop production (CPWF 2003; Ashrafi et al. 2004; Qureshi et al. 2005). The degradation of rangelands is increasing due to overgrazing. These anthropogenic land use changes together with natural factors (low forest canopy covers and high erosion rates) add to the complexity of the hydrology and water resources of the basin considered in the wider spatio-temporal perspective.

The cultivation of food grain crops, mainly wheat, dominates the agricultural land use, besides other crops including fodder, vegetables, maize, sugar beat, pulses and orchards. The dominance of wheat cultivation reflects the country's policy of attaining self-sufficiency in wheat production. The wheat is grown all over the basin, both in rain-fed and irrigated conditions. The land and water productivity of wheat and other major crops is generally low and has a large variation across the basin (Ahmad et al. 2009). For instance, the land productivity of rain-fed wheat is about $1,460 \pm 580$ kg/ha and its water productivity is 0.46 ± 0.22 kg/m^3, indicating considerable scope for improvement. The water scarcity and the high variability of the rainfall within a crop-growth season could be regarded as the major constraints to the crop production under rain-fed conditions, besides other factors such as soil fertility and management-related issues.

The improved availability of water through adopting soil and water conservation techniques and/or by means of providing (supplemental) irrigation could help improve land and water productivity in the rain-fed agricultural systems. These techniques could also contribute to addressing the catchment degradation issue, as they are likely to promote land cover growth and reduce soil erosion. However, a proper understanding their impacts is required for the informed agricultural and water policy formulation process.

The MOE and other institutions have been engaged in the assessment, development and management of the water resources. For instance, a vast network of hydrological stations was established by MOE in the 1950s for monitoring river discharges, climatic variables, sediment yields and water-quality parameters across the whole river system (Figure 6). There were about 50 streamflow gauging stations installed after 1950, but only half of them are used continuously. Consequently, long-term streamflow data are not available for many catchments and the existing records have gaps and quality issues. Filling these data gaps by estimating missing streamflow time series for the poorly gauged catchments is essential for the proper understanding of the spatio-temporal variability of hydrology and water availability in the basin.

JAMAB (1999) undertook assessment of the hydrology and water resources of the basin with the main motivations of developing the available renewable water resources to expand irrigated lands, provide water to increasing populations and industry, control floods and produce hydroelectricity. The basin-level water balance analysis conducted for the hydrological year[4] 1993-1994 shows that, on average, annual precipitation in the basin totals about 25×10^9 m^3/yr. About 66% (16.4×10^9 m^3/yr.) of total precipitation is returned to the atmosphere through ET. The renewable water of the basin accounts for 34% of the total precipitation, equivalent to about 8.6×10^9 m^3/yr., and represents the sum of the amounts of surface water and groundwater. Groundwater exists often in *karsts* (hard rock aquifers) and alluvial aquifers, with the presence of both unconfined and confined conditions. The aquifers have large variations in area and thickness, which have largely been attributed to the tectonic factors, lithology, climate conditions and topography (e.g., JAMAB 1999; 2006a; Tizro et al. 2007). Generally, subsurface water storage in porous aquifers in the northern mountainous regions of the basin is limited to valley floors characterized by relatively large depths, high infiltration rates and good water quality. In the southern arid plains, while the area of porous groundwater bodies increases, the thickness and infiltration decrease and the salinity of groundwater increases.

Out of 7.4×10^9 m^3/yr. of the total streamflows 2.5×10^9 m^3/yr. (or 34%) were diverted to various uses in 1993-94. The direct diversions and pumping from the streams constituted the main mode of water withdrawals in the basin. Groundwater

---

[4] A hydrological year in the Karkheh Basin corresponds to October-September.

contributed about 1.7×10^9 m^3/yr. to agriculture, domestic and industrial uses. Groundwater withdrawals are mainly provided through pumping from deep and shallow aquifers besides natural flow through springs and qanats. The total amount of irrigation water diverted from the surface and subsurface resources in the basin was estimated as 3.9×10^9 m^3/yr., with 63 and 37% contributed through surface and groundwater resources, respectively. Groundwater exploitation is a major source of irrigation in Gamasiab and Qarasou subbasins. Based on the study by JAMAB (1999), the year 1993-94 has been taken as the main reference for the water availability and allocation planning in the Karkheh Basin. The detailed water allocations for different sectors are summarized in Table 1 (JAMAB 1999; 2006b).

Table 1. Current and planned water allocations in the Karkheh Basin, Iran.

Sectors	Water allocation in different years (10^6 m^3/yr.)					
	2001	2006	2011	2016	2021	2025
Rural areas	59	62	66	69	70	67
Urban areas	203	231	242	259	278	295
Mining	0	1	1	1	2	2
Industry	23	30	57	76	93	113
Agriculture	4,149	6,879	6,814	7,135	7,476	7,416
Fish farming	14	119	249	379	477	510
Environment	500	500	500	500	500	500
Total	4,949	7,822	7,929	8,419	8,896	8,902

Notes: The water sources are surface water, groundwater and reservoirs. Data source: JAMAB 1999; 2006b.

The Karkheh Basin remained unregulated by large dams before the completion of the Karkheh Dam in 2001 (details on dams can be found at: http://daminfo.wrm.ir/dam-secondary-fa.html). The Karkheh Dam, having a designed storage capacity of about 7.5×10^9 m^3 (and live storage capacity of about 4.7×10^9 m^3), is a multipurpose dam aimed at providing irrigation water to about 350,000 ha in the Khuzestan plains (in the Lower Karkheh Region) besides the other objectives of hydropower generation and flood control. The various dams and irrigation schemes are currently under construction/study, most notably the construction of another large multipurpose dam, namely Saymareh Dam on the Saymareh River. These massive water works are turning this basin into a largely regulated one.

The ongoing water resources development strategies in the Karkheh Basin have impacted the distribution of water within the basin and will be continuously impacting the basin hydrology. The earlier studies attempted to provide accounts of water resources availability and their development potential but the implications of water development strategies on the basin hydrology and on the different users and uses of water across the basin are not properly investigated. The needs for reserving water to environmental uses have not been adequately assessed in the earlier studies. The upstream-downstream linkages of the water uses are not evaluated and, therefore, are poorly understood and not internalized in the water policies. There is a lack of understanding of the realities of basin hydrology and linkages with water management at the river-basin scale. With such information gaps, a sound knowledge of basin hydrology is essential for effective water development policies so that their negative impacts on different uses and users can be avoided, minimized or mitigated. Therefore, there is a dire need for increasing the knowledge and understanding of basin hydrology in view of the changing phases of water management in the Karkheh Basin. A sound knowledge of spatio-temporal hydrology is also imperative for addressing the pressing water management issues revealed by close consultations with key stakeholders in the Karkheh Basin. These issues are enumerated below (CPWF 2003, 2005; Ashrafi et al. 2004; Qureshi et al. 2005):

- Improving understanding of the hydrology and water management at the river-basin scale.
- Assessing the impacts of present irrigation development strategies on different users in upstream-downstream locations, and how they are influencing the basin hydrology.
- Assessing the environmental water demands in the basin at different scales through developing appropriate assessment methodologies.
- Minimizing land erosion and reducing sedimentation yields to the Karkheh Dam by exploring and implementing better catchment management practices and reversing land degradation caused by different reasons such as overgrazing and increasing agricultural area.
- Managing salinity and waterlogging in the lower parts of the basin.
- Improving the productivity of agricultural water use in irrigated, rain-fed and pastoral systems.
- Finding how water and poverty in the basin are interlinked and determining the potential water-related interventions that can lead to poverty reduction in the basin.

1.3. Research Framework

1.3.1. Research motivation

The role of hydrological analysis remains pivotal in formulating policies and strategies for water resources development and management when stakeholders require more precise assessment of the state of their water resources for making tough water allocation decisions for highly competing water needs such as agriculture, environment and other uses. This will be intensified due to increased stress on the water resources as a result of global changes (climate change, land use change, escalating population and food demands, etc.). The hydrological analysis that can provide a reliable assessment of the state of the water resources, and is able to integrate cause(s) and effect(s) of relationships of natural and human-induced changes on hydrology and water resources across multiple spatio-temporal scales and among multiple users in a river basin remains highly imperative in making sustainable water-related decisions.

1.3.2. Research objectives and questions

The main objective of this research is to provide a hydrology-based assessment of surface water resources of the Karkheh Basin, and study its continuum of variability and change at different spatio-temporal scales.

The specific research questions are as follows:

- What is the state of spatio-temporal variability of surface water hydrology and water availability?
- What are the major natural and anthropogenic factors governing the streamflow regime?
- What are the main features of the natural streamflow variability including both high- and low-flow regimes?
- What is the nature of the observed trends in streamflow (if any) and how are the observed trends associated with climate?
- How can scientifically sound and reliable assessments of rainfall-runoff relationships be made using hydrological models with limited amount of data?
- How can regionalization procedures contribute to catchment modeling under a data-limited environment?
- What are the impacts of developing rain-fed agriculture on downstream flows under different scenarios?

1.3.3. Contribution of the proposed research

This research contributes to Basin Focal Project (BFP) for Karkheh Basin. The BFPs were the initiative of Challenge Program on Water and Food (CPWF) in order to strengthen the basin focus of the program. The main goal of the BFPs were to provide a more comprehensive and integrated understanding of the water, food and environment issues in a basin; and to understand the extent and nature of poverty within each selected basin and determine where water related constraints are both a major determinate of poverty factor and where those constraints can be addressed (CPWF 2005).

The IWMI executed Karkheh BFP in collaboration with several Iranian partners during 2005-2009. The project followed the new IWMI research framework, which focused on analysing water availability, mapping water productivity, mapping water poverty, analysing high potential interventions and assessing impacts (IWMI 2005).

The IWMI research framework and Karkheh BFP research methodology were underpinned by the interdisciplinary knowledge and research/evaluation methodologies for which hydrology and water resources assessments were the important components. This PhD research contributes directly to improve understanding of the Karkheh Basin hydrology and water availability. In general, this research contributes in improving understanding of basin scale hydrological processes exhibited in macro scale semi-arid basin which is quite diverse in hydro-climatic features and is data scarce. The knowledge generated by this study is helpful to improve understanding of spatio-temporal variability of the basin hydrology and its use in the sustainable management of water resources in a river basin context for the Karkheh Basin, and similar regions of Iran and elsewhere.

2.   MATERIALS AND METHODS

2.1.   Methodological Framework

The methodological framework followed in this study is schematised in Figure 9. The spatio-temporal details of analysis depend on the specific research issue, application of a particular method and data availability. These details are specified in the relevant chapters of this thesis. The hydrological variability and water availability were investigated using various state-of-the-art methods of hydrological analysis (termed as system investigation). The hydrological modeling was carried out to understand hydrological process and their variability and to test the impact of the water management interventions. The hydrological analysis and modeling consequently provide a sound scientific basis for guiding the water resources management in the river-basin context. A brief description of the methods used in this study is presented below.

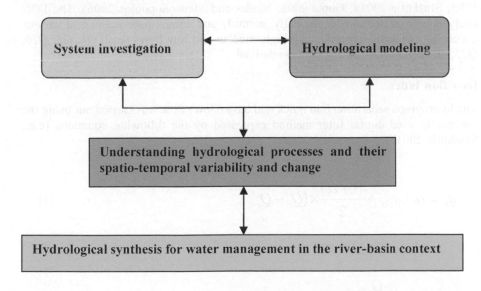

Figure 9. Methodological framework followed in this research study

2.1.2. System investigation

The following methods were used in the system investigation activities.

Measure of central tendency and dispersion

The statistical measures were calculated to understand the central tendency of the streamflow time series. For this the arithmetic means were estimated for monthly and annual flows. Since in semi-arid and arid river basins, like the Karkheh Basin, the arithmetic mean may be biased by a relatively small number of extreme values, median statistics were also computed to get a better understanding of the average conditions. The range of variability was measured by estimating the dispersion in the data by computing the standard deviation and the coefficient of variation (CV).

Flow duration analysis

The flow duration curve (FDC) is a widely used measure in water resources assessment and management for the investigation of water availability for designing hydropower or irrigation schemes, streamflow requirements for riverine ecosystems, etc. The FDC is a cumulative distribution of flows at a site showing the flow assurance of how often any flow is equaled or exceeded. The details on the concept and applications can be found in literature (Linsley et al. 1949; Vogel and Fennessey 1995; Smakhtin 2001a; Gupta 2008; Niadas and Mentzelopoulos 2008). The FDC analysis was carried out for the daily, monthly and annual time scales and various exceeding percentiles representing high, median and low flows (e.g., Q1, Q5, Q10, Q25, Q50, Q75, Q90, and Q95) were derived.

Base flow index

The hydrograph separation into quick and base (slow) flow was carried out using the commonly used digital filter method expressed by the following equations (e.g., Smakhtin 2001b):

$$q_t = \alpha \times q_{t-1} + \frac{(1+\alpha)}{2} \times (Q_t - Q_{t-1})$$ (1)

$$Q_{baseflow} = Q_t - q_t$$ (2)

$$BFI = \frac{Q_{baseflow}}{Q_{total}}$$ (3)

Here, q_t is the filtered direct runoff at time step t (m³/s); q_{t-1} is the filtered direct runoff at time step t-1 (m³/s); α is the filter parameter (-); Q_t is the total runoff at time step t (m³/s); and Q_{t-1} is the total runoff at time step t-1 (m³/s). Then the base flow ($Q_{baseflow}$) is estimated as the difference of Q_t and q_t (equation 2). The Web-based Hydrograph Analysis Tool (WHAT) was used to do the calculations (Lim et al. 2005; http://cobweb.ecn.purdue.edu/~what/). The value of α was set to 0.995 after Smakhtin (2001b) for all of the investigated gauging stations. The main purpose of this exercise was to estimate the Base flow Index (BFI) which is the ratio of the base flow to the total streamflow (equation 3). The BFI estimates were used to characterize the base flow contribution to the streamflows and as well as its spatio-temporal variability. Further details on base flow analysis and some of its applications can be found at Nathan and McMahon (1990a), Arnold and Allen (1999), Larocque et al. (2010) and Welderufael and Woyessa (2010).

Water accounting

The water accounting framework developed by Molden and Sakthivadivel (1999) was applied for the basin-level water accounting. This framework provides a unique way of distinguishing different water use categories such as net inflow, process depletion, non-process depletion, committed water and uncommitted outflows. The key terms of the water accounting methodology, used in this study are defined below. The details can be found in Molden (1997) and Molden and Sakhtivadivel (1999).

- Gross inflow: the total amount of inflow crossing the boundaries of a domain.
- Net inflow: the gross inflow less the change in storage over the time period of interest within the domain. Net inflow is larger than gross inflow when water is removed from storage.
- Process depletion: that amount of water diverted and depleted (or consumptively used) to produce an intended good.
- Non-process depletion: depletion of water by uses other than the process that the diversion was intended for.
- Committed water: the part of outflow that is reserved for other uses such as the environment.
- Uncommitted outflow: outflow from the domain that is in excess of requirements for downstream uses.
- Available water: the amount of water available for a service or use, which is equal to the inflow less the committed water.

Trend and correlation analysis

Trends in the long-term streamflow and climatic data were examined using the Spearman's Rank (SR) test (e.g., McCuen 2003; Yue et al. 2002). The SR test is a nonparametric rank-order test. Given a sample data set $\{X_i, i = 1, 2,..., n\}$, the null hypothesis H_0 of the SR test is that all the X_is are independent and identically distributed. The alternative hypothesis is that X_i increases or decreases with i, so that, consequently, a trend exists. The calculation of the SR statistics R_{sp} requires that the original observations X_is are transformed to ranks k_is by arranging them in the increasing order of magnitude and computed the quantity D_i as $D_i = k_i - i$, where i ranges from 1 to n and n is the number of observations. The R_{sp} and test statistics t were calculated using the following equations:

$$R_{sp} = 1 - \frac{6 \sum_{i=1}^{n} D_i^2}{n(n^2 - 1)} \qquad (4)$$

$$t = R_{sp} \sqrt{\frac{n-2}{1 - R_{sp}^2}} \qquad (5)$$

If the computed t value lies within the desired confidence limits, we can conclude that there is no trend in the series. We used a 90% confidence interval for the evaluating presence or absence of trends. The water limited semi-arid to arid environments, like the Karkheh Basin, are sensitive to changes; therefore, trends that are significant at the 90% level could have quite serious implications. The relationship between streamflow and climatic variables was studied by performing a correlation analysis among them. For this purpose Pearson correlation coefficient, r, was estimated (e.g., McCuen 2003). It is pertinent to note that defining the significance of r values varies with the number of observations and selecting the confidence bound, i.e., in the case of 40 observations, where the values outside the range of ± 0.304 are defined as significant at the 95% confidence interval. However, there is no strict approach for the interpretation of the correlation values and it largely depends on the context and purposes. In this study, the variables were considered having a good correlation if the r values fall outside the critical range of ± 0.304.

Serial correlation

Before applying the trend test, the studied data series were checked for the presence of serial correlation. Previous studies have shown that the existence of serial correlation can complicate the detection and evaluation of trends when applying a nonparametric trend test and, thus, may have strong influence on the null hypothesis about the presence of trends (e.g., von Storch and Navarra 1995; Yue and Wang 2002). The widely used method, termed as "pre-whitening," is used to remove the serial correlation, if present, from the data series before examining the trends. The pre-whitening approach involves calculating the serial correlation and removing the correlation if the calculated serial correlation is significant at the 95% confidence interval (e.g., Douglas et al. 2000; Yue and Wang 2002). The following equation was applied for this purpose:

$$Y_t = X_t - r_1 X_{t-1} \tag{6}$$

where, Y_t is the pre-whitened series value for time interval t, X_t is the original time series value for time interval t, and r_1 is the estimated first serial correlation coefficient. Data were normalized before pre-whitening was carried out, by subtracting the mean and dividing the result by the standard deviation.

In this study, most of the studied variables did not show significant serial correlation. However, when a significant serial correlation was noted, the trend results for that particular case were mentioned for the pre-whitened data.

2.1.3. Hydrological modeling

The two hydrological models Hydrologiska Byråns Vattenbalansavdelning (HBV) and Soil Water Assessment Tool (SWAT) were used. A brief description of these models is given below.

The HBV model

The HBV model was used for regionalization purpose to estimate time series of streamflow at poorly gauged sites. This model was selected for the following reasons: a) its model structure is simple but flexible and can be adapted to local conditions. For instance, a catchment can be subdivided into different elevation and vegetation zones, which fact was important to model catchments in the mountainous Karkheh Basin, b) it is not very data-intensive and most of the data needed are readily available, c) it has been widely used worldwide, in particular in snow-influence climates, but recent studies demonstrate its applicability in semi-arid environments too (e.g., Lidén and Harlin 2000; Love et al. 2010), and d) a number of studies have demonstrated its suitability in regionalization studies (e.g., Seibert 1999; Merz and Blöschl 2004; Götzinger and Bárdossy 2007).

The HBV model (Bergström 1992) is a conceptual rainfall-runoff model which simulates daily discharge using as input variables daily rainfall, temperature and daily or monthly estimates of reference evapotranspiration (ETo). The model consists of different routines representing the snow accumulation and snowmelt by a degree-day method, recharge and actual ET as functions of the actual water storage in a soil box, runoff generation by two linear reservoirs with three possible outlets (i.e., runoff components), and channel routing by a simple triangular weighting function. Further descriptions of the model can be found elsewhere (Bergström 1992; Seibert 1999; 2002; Uhlenbrook et al. 1999). The version of the model used in this study, "HBV light" (Seibert 2002), corresponds to the version HBV-6 described by Bergström (1992) with only two slight changes. Instead of starting the simulation with some user-defined initial state values, this version uses a warming-up period during which the state variables evolve from standard initial values to more appropriate values for the given hydro-meteorological conditions. The length of the warming-up period generally depend on the catchment response characteristics and length of the available data. For this modeling study, the visual inspection of the observed and simulated runoff for the study catchments revealed that a warm-up period should be more than 6 months, and after this period observed and simulated runoff starts matching satisfactorily. Furthermore, the restriction that only integer values are allowed for the routing parameter MAXBAS has been removed, which enables a somewhat more realistic parameterization of the runoff routing processes. The function of MAXBAS is to distribute runoff generated during a time period into the following time steps. In the original version of the HBV model (Bergström 1992) computations in both the snow and soil routine are performed individually for each elevation zone before the groundwater recharge of all zones is lumped in the response routine. In the model version used in this study, the upper box in the response function is treated individually for each elevation zone in addition to the separate computations in the snow and soil routines. This version is considered more logical than the standard HBV versions, especially for use in a mountainous area like the Karkheh Basin.

The SWAT model

The Soil Water Assessment Tool (SWAT) was used for a detailed investigation of hydrological processes and water resources variability and to assess the impact of various water management interventions. This model was selected because: a) it possesses adequate representation of physical processes governing hydrology and is particularly suitable for application to large river basins, b) it is well suited to the proposed research questions on understanding the hydrological processes with limited amount of data, c) it provides a wide range of options for testing "what if" scenarios related to agricultural water management, climate and land use changes etc., and d) it is freely available.

SWAT is a widely used process-based semi-distributed catchment model developed by the Agricultural Research Service of the United States Department of Agriculture (USDA) over the last 30 years and is available free of charge as a public domain model (Arnold et al. 1998; Srinivasan et al. 1998; Arnold and Fohrer 2005; Neitsch et al. 2005; Gassman et al. 2007). SWAT is developed to predict the impact of land and water management practices and of climate change on water, sediment and agricultural chemical yields in large complex watersheds with varying soils, land use and management conditions over long periods of time. It has gained international acceptance as an interdisciplinary tool suitable for applications in large river basins with varying degree of biophysical, climatic and water management settings (e.g., Gassman et al., 2007). The model has been widely applied throughout the world for dealing with a wide range of issues related to hydrology, water management, climate change impacts, land use impacts, best management practices, conservation agriculture, sedimentation and water quality etc. (e.g., Weber et al. 2001; van Griensven and Bauwens 2003; Arnold and Fohrer 2005; Chaplot et al. 2005; Jayakrishnan et al. 2005;Vandenberghe et al. 2005; Faramarzi et al. 2009,

Githu et al. 2009a, 2009b). A comprehensive review on the SWAT model and its applications can be found in Gassman et al. 2007. A brief description of the model is presented here. The detailed theoretical documentation can be found in Neitsch et al. 2005.

In the SWAT model, a river basin is subdivided into a number of subcatchments, each subcatchment consisting of at least one representative stream. The subcatchments are further divided into hydrologic response units (HRUs), which are lumped land areas within the catchment comprising unique land cover, soil, and slope combinations. The hydrology in SWAT is divided into two major divisions, the first being the land phase of the hydrologic cycle, which controls the amount of water, sediment, nutrient and pesticide loadings to the main channel in each catchment and the second being the water routing phase of the hydrologic cycle, which can be defined as the movement of water, sediments, nutrients, etc., through the channel network.

The hydrologic cycle as simulated by the SWAT model is based on the following water balance equation:

$$ SW_t = SW_o + \sum_{i=1}^{t} \left(R_{day} - Q_{surf} - E_a - W_{seep} - Q_{gw} \right) \tag{7} $$

where, SW_t is the final soil water content (mm), SW_0 is the initial soil water content on day i (mm), t is the time (days), R_{day} is the amount of precipitation on day i (mm), Q_{surf} is the amount of surface runoff on day i (mm), E_a is the amount of evapotranspiration (ET) on day i (mm), W_{seep} is the amount of water entering the vadose zone from the soil profile on day i (mm), and Q_{gw} is the amount of return flow on day i (mm). The surface runoff volume is calculated by using the Soil Conservation Service (SCS) curve number equation. Potential evapotranspiration can be estimated by one of the three methods: Penman–Monteith, Priestly and Taylor or Hargreaves method. The actual ET is estimated on the basis of simulated plant growth and soil water availability. The model calculates percolation when the soil-water content exceeds the soil-field capacity and determines the amount of water moving from one soil layer to the next by using a storage routing method. In each subcatchment, the SWAT model simulates two groundwater aquifers: a shallow aquifer that contributes to streamflow and a deeper aquifer that does not add to streamflow within the modeled subcatchment. Streamflow is routed by using either the variable storage or the Muskingum routing method.

There are numerous other processes represented in SWAT, such as water balance for lakes/ponds/reservoirs, sediment erosion and sediment transport processes, industrial and municipal pollution added through point sources, processes related to transformation and movement of nitrogen and phosphorus.

The SWAT 2005 version is well linked to geographic information system (GIS), ARC-GIS, which have further enhanced its abilities to deal with spatial information for management, query, visualization and analysis. The model also has added features for auto-calibration, sensitivity and uncertainty analysis. In this study, the SWAT 2005 modeling system, version ARCSWAT 2.0 (Winchell et al. 2008), was used.

Statistics used in the performance evaluation

The model performance was assessed by using three performance measures; Nash-Sutcliffe Efficiency (*NSE*) (Nash and Sutcliffe 1970), Coefficient of Determination (R^2) and the mean annual volume balance (VB). These criteria are most widely recommended and commonly used in hydrological modeling (e.g., ASCE 1993; Gupta et al. 2009). The VB is estimated as a percentage difference between the observed and simulated mean annual runoff. The equations for estimating *NSE* and R^2 are given below:

$$NSE = 1 - \frac{\sum (Q_{obs} - Q_{sim})^2}{\sum (Q_{obs} - \overline{Q_{obs}})^2} \qquad (8)$$

$$R^2 = \frac{\left(\sum (Q_{obs} - \overline{Q_{obs}})(Q_{sim} - \overline{Q_{sim}})\right)^2}{\sum (Q_{obs} - \overline{Q_{obs}})^2 \sum (Q_{sim} - \overline{Q_{sim}})^2} \qquad (9)$$

where: Q_{obs} and Q_{sim} refer to the observed and simulated discharges, respectively. and $\overline{Q_{obs}}$ and $\overline{Q_{sim}}$ refer to the mean of the observed and simulated discharges, respectively. The observed and simulated streamflows and their means will have same units, i.e., expressed as m^3/s in this study.

2.2.   Data Collection

The International Water Management Institute (IWMI: http://www.iwmi.cgiar.org/) in collaboration with local partners has conducted research through the Karkheh Basin Focal Project (BFP) in Iran to address some of the issues and challenges discussed in the previous chapter. This project was funded through the Challenge Program on Water and Food (CPWF: http://www.waterandfood.org/). The main aims of the Karkheh BFP were to provide a comprehensive and integrated understanding of the water, food and environment issues in the river-basin context. Various data sets were accumulated from global and local sources under this Project. All of these data sets were managed under the Integrated Database Information

System (IDIS) - a database management project of IWMI and CPWF based in Colombo, Sri Lanka. This research study was an integral component of the Karkheh BFP, and thereby, a number of data sets collected through primary and secondary sources under this project were used in this research. The major data sets used in this study are listed in Table 2. Further details are provided in the relevant chapters.

The quality of collected data sets was checked in a number of ways, mainly depending on the type of the data set and perceived uncertainties. For instance, land use map prepared under Karkheh BFP by IWMI was considered of reasonably good quality, because it used sound scientific basis in its preparation and was extensively validated through field observations. The quality of hydro-climatic data sets was examined by visual inspection of the tabular data and their graphical presentations. Moreover, double mass analysis was used to check the consistency of the hydrological time series (Change and Lee 1974; Linsley et al. 1982). In the double mass analysis, if a linear relationship is found between an individual station and the mean of its neighbours, or the remainder of the set within a basin, then it is inferred that the data series has been recorded consistently over its history. The results of the double mass analysis applied on the time series data of the selected flow gauging stations found no deviations from the corresponding linear plots, indicating that records were consistent.

Table 2. An overview of main data sets used in this study.

Category	Data	Data source
Hydrology	River discharge	MOE, Iran
Climate	Precipitation, temperature, relative humidity, sunshine hours, wind speed	Meteorological organization, Iran, MOE, Iran
Topography	Digital elevation model (DEM)	Shuttle Radar Topography Mission (SRTM) of USGS
Soils	Digital map of the soils and soil properties	Soil and Water Research Institute (SWRI), Iran, other relevant departments, and FAO 1995 soil map of the world
Land use	Land use map	IWMI Karkheh BFP
Irrigation	Irrigation diversions and on-farm irrigation practices	IWMI Karkheh BFP and relevant Iranian sources
Agriculture	Crops, yields, agronomic practices, agricultural statistics	IWMI Karkheh BFP and relevant Iranian sources

3. STREAMFLOW VARIABILITY AND WATER ALLOCATION PLANNING[5]

3.1 Introduction

Arid, semi-arid and subhumid regions are called water limited environments and occupy about half of the global land area (Parsons and Abrahams 1994). Changes in water availability can have serious repercussions on the sustainability of these sensitive environments. The pressure on water and other natural resources is increasing in these areas as demands for water for human uses are growing rapidly (e.g., Newman et al. 2006). For instance, in the dryland Mediterranean regions, large increases in population, development of irrigated agriculture and rise of living standards have drastically increased the water use and in many basins future needs are hard to satisfy as many aquifers are already overexploited and surface water resources are endangered (Cudennec et al. 2007). Southern Africa faces similar challenges (e.g., van der Zaag 2005). The expected regional climate change (Christensen et al. 2007) poses yet another dangerous alteration of the hydrological regimes in these regions. This will also cause change in the water demand pattern, with an expected two-thirds of the world facing an increase in irrigation demand (Döll 2002).

The semi-arid to arid Karkheh Basin has a fragile balance between environmental and human uses of natural resources and demands for water are increasing and sustainable management of water resources has become an important issue. The main challenges related to land and water resources are land degradation, soil erosion, low water and land productivity, groundwater depletion and growing competition for water among upstream and downstream areas and among different sectors of water use such as irrigation, domestic, hydropower and environment (CPWF 2005). In this river basin, massive irrigation development is on the way, but the knowledge and understanding of basin hydrology (including the spatio-temporal water balance variations) and impacts of these developments on other users and water uses across the basin are patchy.

Quantitative and holistic knowledge of basin hydrology becomes essential as water-management needs become complex. Molle et al. (2004) concluded that, as water demands increase and more and more water is allocated to different uses, the management of water resources becomes increasingly complex due to the huge

[5] This chapter is mainly based on, but not limited to, the paper Analysing streamflow variability and water allocation for sustainable management of water resources in the semi-arid Karkheh River Basin, Iran by Masih, I., Ahmad, M.D., Uhlenbrook, S., Turral, H. and Karimi, P. 2009. *Physics and Chemistry of the Earth* 34 (4-5): 329-340.

number of interacting factors such as upstream-downstream impacts, increasing impacts on the environment and changes in *de facto* water rights. They have argued that under such conditions, increasing the knowledge of the basin hydrology is essential for constructing a sound and sustainable water management. A sound knowledge of basin hydrology is essential for effective water allocation policies so that negative impacts can be avoided, minimized or mitigated (Green and Hamilton 2000). Hydrological analysis provides the basis for detailed accounting of water use and productivity (Molden and Sakthivadivel 1999). It is a basic requirement for water resources development and management evaluations and decision making related to a) assessing water availability, b) understanding the balance between the actual use resource availability, c) improving water allocation decisions, d) monitoring the performance of water use, and e) formulating environmental flow requirements and working out ecosystem restoration strategies.

This chapter provides a comprehensive analysis of spatio-temporal variability of the surface water hydrology over the period of 1961 to 2001 in the Karkheh Basin. Additionally, basin-level water accounts are evaluated for the year 1993-94 and challenges for sustainable management of water resources are highlighted.

3.2.   Data and Methods

For this study seven streamflow gauging stations of the main rivers (as shown in Figure 6 and Table 3) were selected. The rationale for selecting these stations includes their geographical importance, availability of consistent length and quality of the records. Out of the seven stations, three stations namely Pole Chehre at the Gamasiab River, Ghore Baghestan at the Qarasou River and Pole Dokhtar at the Kashkan River are located at the outlet of their respective sub-basins. The Holilan at the Saymareh River represents the combined effect of the hydrologic characteristics of the upstream sub-basins Gamasiab and Qarasou. The Jelogir at the Karkheh River is located upstream of Karkheh dam and the Paye Pole station located downstream of the Karkheh dam is important for water supplies for hydropower and downstream flows for irrigation and environment. The Hamedieh station is the last gauging station before the Karkheh River routes towards Hoor-Al-Azim swamp and hence is important for environmental flows further downstream, i.e., towards Hoor-Al-Azim swamp.

The analysis was conducted using daily streamflow data for the period 1961-2001. This data set was used for the analysis of central tendency and dispersion, flow duration analysis, base flow separation and water accounting. Basin-level accounting of water use was carried out using available data for the year 1993-94. A description of these methods is provided in section 2.2 of this thesis.

The data on water accounting components were accumulated from the study of JAMAB (1999) who conducted comprehensive water balance investigations for the hydrological year 1993-94. A brief description of their methodology is provided here; further details can be found in JAMAB 1999. The estimates were mainly based on the data related to climate, runoff and water uses. For the purpose of a detailed water balance analysis, the basin was divided into 47 subcatchments, and the results were aggregated to the basin level. The inflow components of the water accounting were composed of precipitation, inflow from outside of the basin and changes in surface and subsurface storage. The precipitation data were based on 61 climatic stations distributed in or close to the basin. Changes in the subsurface storage were estimated based on the groundwater measurements related to changes in the water level, specific yield and domain area. Since there was no major storage dam in the basin during 1993-94, the surface storage was considered zero. Inflow from outside of the basin was zero, as no water was diverted to the Karkheh from outside of the basin. The actual *ET* was estimated through empirical equations calibrated for selected locations in the basin where detailed data on climate and water balance were available. The actual *ET* from diversions for agricultural and other purposes was estimated as the difference between the total abstraction and return flows. The return flows were estimated for industrial, domestic and agricultural sectors for each of the 47 subcatchments, and were based on field observations. The outflow from the basin was composed of outflow from rivers, drains and subsurface outflow. The subsurface outflow was regarded as zero, whereas, outflow from rivers and drains was based on the observed records. The data on committed and uncommitted water were not available, but are necessary to complete the water accounting exercise. We estimated the committed water to the range of 10 to 50% of the available annual streamflows. This estimation was based on the study of Tennant (1976) who suggested that allocating 50% of the available streamflows to the river ecosystems can maintain healthy ecosystems whereas the minimum flows should be 10%, though the ecosystem degradation will be inevitable at this level of allocations.

Table 3. Geographical characteristics of the selected river stations.

Name of river	Name of station	Longitude (degrees East)	Latitude (degrees North)	Elevation (masl)	Drainage area (km²)
Gamasiab	Pole Chehre	47.43	34.33	1,280	10,860
Qarasou	Ghore Baghestan	47.25	34.23	1,268	5,370
Saymareh	Holilan	47.25	33.73	1,000	20,863
Kashkan	Pole Dokhtar	47.72	33.17	650	9,140
Karkheh	Jelogir	47.80	32.97	450	39,940
Karkheh	Paye Pole	48.15	32.42	125	42,620
Karkheh	Hamedieh	48.43	31.50	20	46,121

Data source: MOE, Iran.

3.3.   Results and Discussion

3.3.1. Spatial and temporal variability of daily streamflow regimes

The daily streamflows show large variability within a year and between years, as exemplified in Figures 10 and 11. However, the general temporal patterns are quite synchronized, with rising and falling limbs of the hydrographs most often corresponding to similar timings, when different streams or flow behaviors at different locations on a single river are compared. For tributary rivers (Figure 10), the highest streamflow is observed at Holilan, which aggregates the streamflows coming downstream from Pole Chehre and Ghore Baghestan. It is pertinent to note that despite higher drainage area, Pole Chehre has lower streamflows than Pole Dokhtar. This could be mainly attributed to comparatively lower precipitation and higher agricultural water use in the case of Pole Chehre. For the Karkheh River at Jelogir, Paye Pole and Hamedieh (Figure 10), the flows are very similar to one another. This could be attributed to the fact that most of the streamflows are generated before Jelogir. Although there are some abstractions used for irrigation downstream of Jelogir and Paye Pole, these were not high enough to cause major differences in the flow regimes, mainly due the absence of any major water infrastructural project (e.g., large dams) during the period under study. However, it could be anticipated that due to operations of the Karkheh Dam (operational since 2001) and new irrigation schemes, the streamflow regimes of these three stations would be markedly different from one another.

The high flow events are mainly concentrated in the months from November to May, particularly in February and March. The duration of these events varies largely depending on the precipitation timing and snowmelt conditions. Generally, high flow events of small duration (1-5 days) occur due to high rainfall events, but the high flows prevailing for a few weeks to a couple of months, mainly observed from February to May, are caused by the snowmelt and the combined effects of snowmelt and rainfall. The low flow regime is marked from June to September. The high spatial and intra-annual variability in the streamflows is mainly governed by the seasonality of climate and by factors such as land use, geology, soils and topography. Most of the precipitation occurs during winter, both in terms of rain and snow, and in spring mainly as rain. The snowfall occurs from December to March, with the highest amounts in January and February. The amount of snow varies within the basin, with upper parts receiving more than the middle parts and no snowfall in the lower arid region. A recent study on snow cover in the Zagros mountains by Saghafian and Davatalab (2007) showed that the snow water equivalent to the mountainous parts of the Karkheh Basin is about 75 mm/yr., which is about 17% of the long-term mean annual precipitation in the basin. Moreover, the amount and distribution of snow are strongly influenced by elevation varying from 44 mm/yr. in locations less than 1,500 masl to 245 mm/yr. in locations over 3,500 masl.

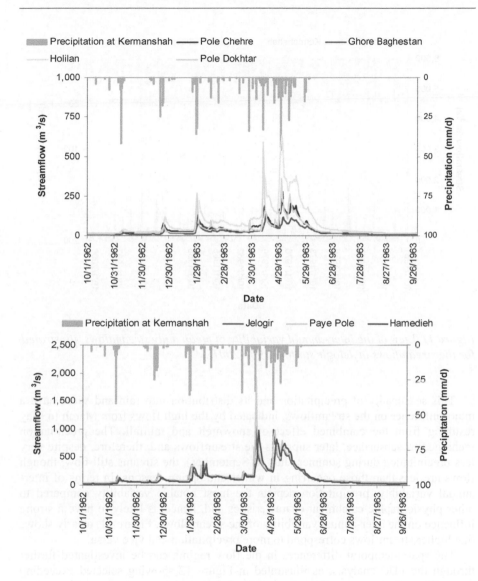

Figure 10.   Intra-annual variability of mean daily streamflows, illustrated for the data of the hydrological year 1962-63.

(Note: The precipitation at Kermanshah was used as an example in this figure and in few other figures in this chapter. The main aim is to illustrate the general pattern of the precipitation in the study basin. The major reason for selecting this station was due to availability of good quality long-term daily data series at this site. It is important to note that the precipitation values observed at this station are not fully representative of the amount and distribution of precipitation in the whole basin. More details on precipitation dynamics and their influence on runoff are given in chapter 4 and 6.)

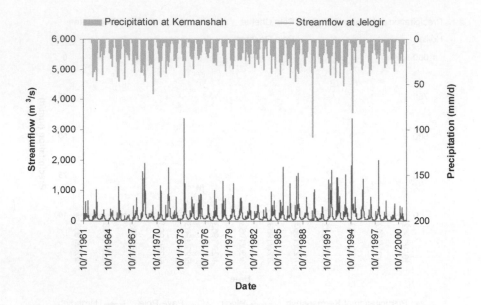

Figure 11. View of the inter-annual variability of mean daily streamflows, illustrated for the streamflows at Jelogir station (1961-2001).

This seasonality of precipitation and its distribution into rain and snow have a major influence on the streamflows, indicated by the high flows from March to May resulting from the combined effect of snowmelt and rainfall. The precipitation recharged to subsurface, later supports the streamflows and, therefore, despite very less precipitation during summer (July to September), the streams still flow, though flows are less than those occurring in winter and spring seasons. In terms of inter-annual variability, precipitation depicts the most notable variability compared to other physiographic catchment characteristics and, hence, is likely to have a strong influence on the inter-annual variability of the streamflows. Figure 11, clearly shows that higher streamflows correspond to more precipitation and vice versa.

The spatio-temporal differences in the flow regime can be investigated further through the FDC analysis, as illustrated in Figure 12 showing selected exceeding percentiles of the streamflows normalized by the drainage area. The actual streamflows are also provided in Table 4. It is worth noting from Figure 12 that the FDC of Pole Dokhtar plots higher compared to all other stations, even to those with higher streamflows (e.g., Paye Pole), whereas the FDC of Pole Chehre plots the lowest. This is attributed to higher specific runoff for Pole Dokhtar than for Pole Chehre and other locations. Furthermore, the steeper slopes of FDCs observed at Pole Chehre and Ghore Baghestan indicate less stable flow regimes having lower proportions of the base flow than at Pole Dokhtar. The flow regime at Holilan and Jelogir demonstrates the net effects of the upstream tributaries.

The base flow constitutes a quite significant part of the total streamflows all across the basin, particularly for Pole Dokhtar, Jelogir, Paye Pole and Hamedieh, as indicated in Figure 12 and by the annual Base Flow Index (BFI) values close to 0.5 for these gauging stations (Table 5). These BFI estimates suggest that the role of slow flow part of the hydrograph is quite significant in sustaining the streamflows in middle and lower parts of the basin as compared to upper parts showing lower values of BFI (e.g., Pole Chehre's BFI is 0.36), where quick flow dominates volumetrically the overall flow regime.

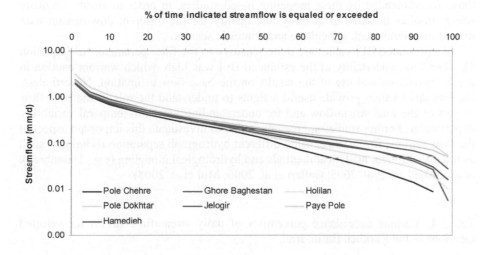

Figure 12. The flow duration curves (FDCs) of selected gauging stations.
(FDC plots are based on selected flow percentiles extracted from the daily streamflow data for the period October 1, 1961 to September 30, 2001.)

The main reasons for the stable base flow regime and a less steeper FDC slope in the case of Pole Dokhtar than in o Pole Chehre and Ghore Baghestan are likely to be the higher precipitation amounts in the middle parts of the basin, higher proportion of forest area which leads to higher infiltration of precipitation that later slowly discharges to rivers via subsurface routes and comparatively fewer irrigated areas in the middle parts of the basin (which mean less water withdrawals from streams and aquifers). The flow regimes of the Karkheh River at Jelogir, Paye Pole and Hamedieh are largely similar to one another, with slightly more stable base flows in the case of Paye Pole. This could be attributed to the presence of the Karkheh Lake just above the Paye Pole station providing some attenuation to the streamflows and then contributing stored water as base flows. The impact of the Karkheh Dam is not evident in this analysis because the dam started operations only in 2001. Evaluating the impact of dams on flow variability was beyond the scope of this research.

However, it is anticipated that the natural flow regime of the Karkheh River below the Karkheh dam would be changed as a result of the reservoir operations. Further details on dams impacts and operational strategies can be found at number of studies conducted in Iran (e.g., Manouchehri and Mahmoodian 2002; Mousavi et al. 2004; Karamouz et al. 2003, 2006, 2008, 2011; Ganji et al. 2007; Kerachian and Karamouz 2007; Zahraie et al. 2008; Zahraie and Hosseini 2009). For instance, the research studies conducted by Karamouz et al. (2006, 2008, 2011) for the Karkheh dam have highlighted the possible conflicts in downstream water availability between various sectors of water use below the Karkheh dam. These investigations emphasized the need of development and adoption of sound reservoir operating policies, such as those recommended by these modeling based studies, in order to ensure adequate water supplies in terms of quantity and quality for the different downstream uses related to environment, agriculture and domestic sectors.

The presented BFI values are quite sensitive to the filter parameter 'α' (equation 1). Therefore, uncertainty of the estimated BFI was high, which warrant caution in the interpretation and use of the results on the base flow estimation. Nevertheless, the estimated values provide useful insights to understand the role of the base flow as part of the total streamflow and for understanding its spatio-temporal variability in particular. Further studies are recommended to investigate this important aspect of the flow regime, for instance, using different hydrograph separation techniques such as tracers, different analytical methods and hydrological modeling (e.g., Uhlenbrook et al. 2002; Lim et al. 2005; Gallart et al. 2006; Mul et al. 2008).

Table 4. Various exceedance percentiles of daily streamflow (m³/s) for selected locations in the Karkheh Basin, Iran.

	Pole Chehre	Ghore Baghestan	Holilan	Pole Dokhtar	Jelogir	Paye Pole	Hamedieh
Q1	245.0	149.0	518.0	325.0	924.0	1,120.0	1,018.0
Q5	126.9	80.8	266.0	171.4	516.0	588.9	563.0
Q10	87.1	55.2	192.0	116.2	365.0	418.0	393.0
Q20	51.9	34.0	113.0	71.8	233.0	265.0	245.0
Q30	34.9	21.9	73.6	48.0	156.0	183.0	173.0
Q40	25.0	15.0	53.9	36.0	118.0	135.0	122.0
Q50	16.4	11.0	35.6	28.6	89.6	101.0	84.0
Q60	8.9	7.9	24.0	22.9	68.8	80.0	61.6
Q70	5.5	5.9	16.0	19.3	54.0	66.5	47.6
Q80	3.5	4.2	11.5	15.5	42.0	52.0	36.0
Q90	1.8	2.5	7.0	12.0	31.0	42.0	24.0
Q95	1.1	1.7	4.7	9.7	24.4	34.9	16.5
Q99	0.0	0.4	2.4	5.6	12.9	25.0	8.7

Note: These exceedance percentiles are extracted from the FDC analysis of the daily data for the period of October 1, 1961 to September 30, 2001.

3.3.2. Spatial and temporal variability of monthly streamflows

Mean monthly discharges at the selected river stations are shown in Figure 13. The hydrograph peaks occur in March and April, roughly one month in lag of precipitation. This could be attributed to contributions of snowmelt in the late winter and early spring seasons as well as contributions of water into streams after passing through different hydrological pathways (such as groundwater). The peak flows are observed in April at all the examined stations whereas minimum flows occur in September. Although most of the discharge takes place in winter (about 41%) and spring (about 39%), all the main tributaries of the Karkheh River have some flow all around the year. The hydrograph separation analysis indicates that the base flow contributions are mainly responsible for keeping the streams flowing during half of the water year, particularly from June through September (Table 5).

The river flows show quite high variability both with respect to space and time, as indicated by high CV (Figure 14). The maximum values of CV are observed for November and it corresponds to river flows at all of the seven selected stations in the basin ranging from 0.96 for Pole Dokhtar to 1.77 for Pole Chehre. Minimum values of CV are observed in February with the spatial variation ranging from 0.44 to 0.53. For rest of the months, the values are in the range of 0.4 to 0.9.

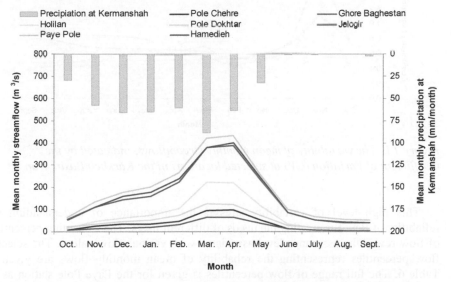

Figure 13. Mean monthly discharge at selected locations in the Karkheh Basin, Iran.

Table 5. Base Flow Index (BFI) for selected locations in the Karkheh Basin, Iran.

	Pole Chehre	Ghore Bagestan	Holilan	Pole Dokhtar	Jelogir	Paye Pole	Hamedieh
Oct.	0.37	0.64	0.55	0.72	0.65	0.69	0.64
Nov.	0.22	0.43	0.34	0.55	0.47	0.49	0.43
Dec.	0.25	0.38	0.31	0.50	0.41	0.42	0.38
Jan.	0.32	0.40	0.36	0.51	0.45	0.45	0.41
Feb.	0.33	0.34	0.33	0.43	0.39	0.41	0.39
Mar.	0.28	0.28	0.28	0.35	0.34	0.38	0.36
Apr.	0.39	0.39	0.40	0.46	0.46	0.49	0.47
May	0.72	0.71	0.70	0.71	0.74	0.78	0.74
June	0.94	0.95	0.94	0.96	0.98	0.97	0.93
July	0.90	0.93	0.98	0.97	0.97	0.94	0.87
Aug.	0.88	0.93	0.97	0.96	0.96	0.90	0.89
Sept.	0.71	0.85	0.86	0.90	0.89	0.87	0.84
Annual	0.36	0.41	0.38	0.49	0.47	0.49	0.46

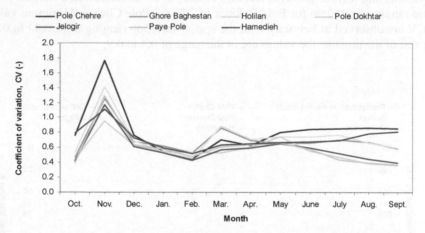

Figure 14. The variability of mean monthly streamflows, indicated by the Coefficient of Variation (CV) at selected locations in the Karkheh Basin, Iran.

This high level of variability stresses the importance of understanding the reliability of flow for meeting the needs of different users. The exceeding percentiles of flow reveal the assurance level associated with various flow values. The selected flow percentiles representing the reliability of mean monthly flows are given in Table 6. The full range of flow percentiles is given for the Paye Pole station as an example (Figure 15). For example, the minimum value of Q_{70} at Paye Pole corresponds to the month of September, indicating that the mean monthly streamflow of 41.4 m³/s is likely to be available for 28 out of 40 months according to the study period (70% of the time or 7 out of 10 months). The maximum mean monthly flow of 285.5 m³/s with a reliability of 70% is available in March at the Paye Pole.

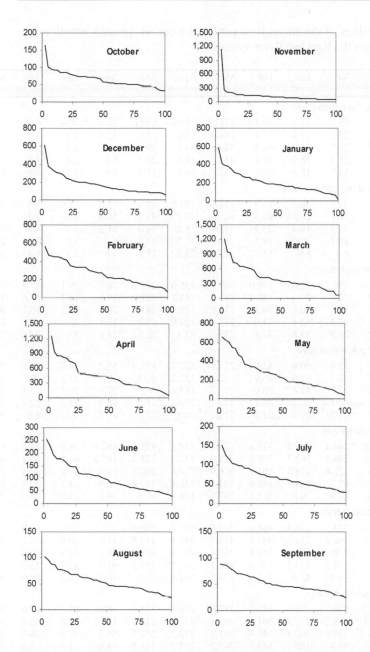

Figure 15. The reliability of the mean monthly surface water availability, indicated by the monthly FDCs at the Paye Pole station at the Karkheh River.
(Note: The x-axis shows percentage of time mean monthly flow was equaled or exceeded, whereas the y-axis shows mean monthly streamflows (m³/s). These exceedance percentiles are extracted from the FDC analysis of the flow data for the period October 1, 1961 to September 30, 2001.)

Table 6. Selected values of the streamflow percentiles (m^3/s) for each month at the studied stations across the Karkheh River system.

Indicator	Oct.	Nov.	Dec.	Jan.	Feb.	Mar.	Apr.	May	June	July	Aug.	Sept.
Pole Chehre station at the Gamasiab River												
Q10	16.0	41.8	55.3	57.3	78.7	171.5	186.9	139.0	24.6	10.4	6.0	5.4
Q30	8.0	26.2	35.2	37.8	57.4	104.4	110.1	61.6	12.4	5.9	4.0	4.0
Q50	5.5	14.9	26.7	31.2	44.8	78.6	84.3	39.6	8.6	4.0	2.3	2.6
Q70	4.0	11.5	21.1	26.5	34.6	63.2	59.9	29.7	5.6	2.4	1.5	1.6
Q90	2.3	6.6	14.3	18.3	23.7	33.2	36.5	11.3	2.9	1.2	0.9	0.7
Ghore Baghestan station at the Qarasou River												
Q10	9.1	20.3	34.2	38.2	55.9	124.0	123.5	80.4	23.6	12.0	9.5	7.0
Q30	5.9	10.5	16.1	26.3	40.5	66.8	77.5	43.7	16.4	8.1	5.5	4.7
Q50	5.2	9.1	13.3	16.4	25.7	55.0	51.2	34.0	11.2	5.9	4.0	3.5
Q70	4.1	6.5	10.7	12.0	19.7	35.9	37.7	21.7	6.9	3.5	2.3	2.3
Q90	2.4	4.3	6.4	8.9	13.7	22.4	23.5	12.5	4.8	2.4	1.4	1.7
Holilan station at the Saymareh River												
Q10	28.7	81.3	101.2	124.8	178.9	395.2	435.6	281.2	76.7	29.1	17.4	15.4
Q30	18.8	46.9	71.8	79.3	127.4	251.2	241.7	127.2	37.6	18.7	12.6	11.9
Q50	15.7	30.8	55.8	60.8	96.1	175.0	179.0	98.6	26.5	15.3	10.2	9.8
Q70	11.7	24.2	42.3	52.4	74.0	133.6	124.6	65.9	15.1	7.9	5.8	5.9
Q90	7.5	14.2	26.0	33.8	47.6	72.9	82.4	32.8	11.0	5.6	3.2	3.8
Pole Dokhtar station at the Kashkan River												
Q10	31.1	53.5	85.6	90.4	129.1	224.5	222.7	151.9	52.7	34.4	26.4	22.5
Q30	21.9	35.5	57.7	57.6	85.9	142.3	153.9	86.2	32.8	23.1	19.5	18.3
Q50	19.2	26.2	37.3	41.4	66.0	110.7	113.1	66.5	27.9	19.1	15.4	14.9
Q70	15.4	20.9	29.1	35.2	48.7	83.3	70.6	41.8	18.3	13.8	12.4	12.3
Q90	11.4	17.2	21.5	26.5	33.8	51.2	48.1	22.3	11.8	9.2	8.6	9.5
Jelogir station at the Karkheh River												
Q10	81.4	153.0	246.4	303.4	375.6	678.2	739.5	495.4	152.1	98.8	71.6	63.3
Q30	64.9	107.2	168.7	162.7	275.6	423.9	432.7	275.4	102.3	61.7	49.4	44.5
Q50	48.7	82.6	126.8	136.8	194.0	326.9	374.5	200.5	76.3	47.4	40.5	38.5
Q70	42.1	69.3	92.7	114.9	168.4	244.3	236.4	137.8	52.9	34.5	30.6	31.1
90	33.5	49.6	67.8	79.8	112.6	162.7	148.8	69.8	34.8	24.0	20.9	21.4
Paye Pole station at the Karkheh River												
Q10	92.1	207.0	323.3	378.5	445.0	729.8	847.8	540.8	178.3	104.2	86.9	79.4
Q30	74.6	143.5	200.2	233.6	334.1	474.5	485.1	314.5	117.7	80.0	63.5	60.9
Q50	58.2	100.1	150.9	178.2	248.8	363.4	406.9	219.0	96.0	61.9	50.8	46.6
Q70	51.3	75.9	102.3	137.5	191.8	285.5	261.8	157.3	61.7	48.9	43.3	41.4
Q90	43.8	57.9	81.4	82.0	114.3	154.7	162.4	91.8	46.4	36.7	32.0	31.4
Hamedieh station at the Karkheh River												
Q10	79.9	183.2	298.3	356.6	442.2	757.9	796.8	531.1	192.6	92.7	67.0	60.2
Q30	59.5	125.5	183.2	214.9	301.3	483.1	444.4	318.8	105.0	57.8	42.8	45.0
Q50	48.0	82.3	128.5	160.1	240.7	331.2	373.2	202.7	65.5	49.6	38.7	36.4
Q70	35.3	61.5	87.4	120.8	163.5	245.2	237.7	148.8	48.6	31.8	27.0	26.1
Q90	19.2	40.5	60.0	61.5	94.3	130.8	143.6	68.8	23.8	19.5	14.3	13.4

3.3.3. Long-term variability in annual surface water availability

The long-term temporal behavior in the annual river flows has similar patterns throughout all the subbasins, where wet and dry years prevail over all areas simultaneously (Figure 16). The annual values of CV fluctuate around 0.47 within a range of 0.41 to 0.54. A comparison of mean and median annual water availability indicates that the mean values are 0-7% higher than the median estimates (Table 7). This exhibits the classic arid and semi-arid hydrology characteristic that the mean is greater than the median but, in this case, not by a large margin at an annual scale (only 4% on average). The maximum flow of 12.59×10^9 m^3/yr. occurred in the wet year of 1968-69 whereas the minimum flow of 1.92×10^9 m^3/yr. corresponds to the drought year of 1999-2000, at the Paye Pole station. In the time period of this analysis, i.e., 1961 to 2001, the severest drought occurred from 1999 to 2001 though the longer-term time series depicts both high and low flow years throughout the study period. During this persistent drought the Gamasiab River ceased to flow during part of the year, indicated by zero flow of 44 to 77 days in a year observed at Pole Chehre but other examined locations recorded some flows.

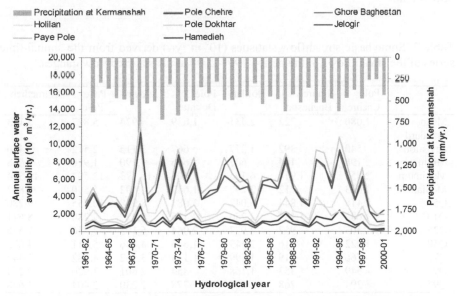

Figure 16. Long-term variability in annual surface water availability across the Karkheh Basin.

These large temporal variations indicate the high level of supply insecurity for current and future increased withdrawals for human uses. The analysis of flow duration curves (Table 7 and Figure 17) provides further insights into the assurance levels of the annual availability of surface water across the Karkheh Basin. For instance, the value of $Q75$ at Paye Pole is 4.08×10^9 m^3/yr., which shows that this much volume of surface water could be available for 75% of the time, i.e., 30 out of 40 years as per duration of the study. Further examination was done to ascertain the assurance levels associated with mean annual water availability. For this, FDC plots were generated, using annual data, (Figure 17) and the exceedance level of means was noted for each station. This analysis indicated that mean annual surface water availability has an assurance level of about 45% at the basin level, ranging from 40% for Pole Chehre to 52% for Pole Dokhtar. This shows that the annual mean is biased towards hydrological years with high values for Pole Chehre (and also for Ghore Baghestan) and, therefore, the median is a better measure of central tendency for these stations. Furthermore, due to the construction of the Karkheh Dam and ongoing irrigation schemes in downstream parts, one can anticipate that, during the below-average/low-flow years, the most serious conflict would concern retention of water in the Karkheh Dam for hydropower generation and reduced supplies to the downstream agricultural users whose situation will be exacerbated by soil salinity problems. This would also be accompanied by the diminished flows to river ecosystem and floodplains as well as to the Hoor-Al-Azim Swamp further downstream.

Table 7. Some basic streamflow statistics (10^6 m^3/yr.) derived from the annual time series of the streamflows of the period 1961-2001 at the selected flow gauges.

	Pole Chehre	Ghore Baghestan	Holilan	Pole Dokhtar	Jelogir	Paye Pole	Hamedieh
Mean	1,080	722	2,431	1,639	4,974	5,827	5,153
Standard deviation	540	392	1,277	667	2,115	2,512	2,476
Minimum	198	104	607	645	1,790	1,916	1,068
Maximum	2,851	1,914	6,193	3,206	10,773	12,594	11,324
Median	1,003	712	2,292	1,637	4,692	5,590	4,852
Q5	2,416	1,844	6,042	3,081	8,958	10,755	9,280
Q10	1,684	1,183	4,250	2,455	8,227	9,280	8,641
Q25	1,303	957	2,977	2,064	6,193	7,756	7,555
Q50	1,022	716	2,343	1,645	4,836	5,651	4,873
Q75	766	419	1,499	1,113	3,562	4,082	3,447
Q90	549	353	1,168	854	2,601	3,020	2,254
Q95	294	268	871	778	2,230	2,404	1,648

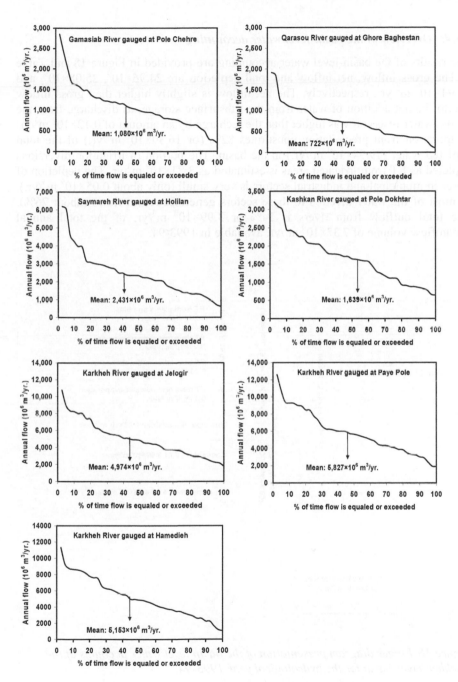

Figure 17. The reliability of the annual surface water availability, indicated by annual FDCs at the selected gauging stations across the Karkheh River system.

3.3.4. Overview of the basin-level water accounting

The results of the basin-level water accounting are provided in Figure 18 and Table 8. The gross inflow, net inflow and total depletion are 24.96×10^9, 25.08×10^9, and 19.94×10^9 m^3/yr., respectively. The net inflow is slightly higher than gross inflow due to the net addition of water from the subsurface storage, as discharge from the groundwater reservoir was higher than the recharge by an amount of 0.12×10^9 m^3/yr. Evaporation from precipitation constitutes 82% (or 16.39×10^9 m^3/yr.) of the total depleted water (19.94×10^9 m^3/yr.) in the basin. The portion of irrigation diversions depleted as ET from irrigated areas is estimated as 3.21×10^9 m^3/yr. The depletion of water in municipal and industrial sectors is very small (only about 0.05×10^9 m^3/yr.), as most of the water diverted to these sectors generates return flows (about 76%). The total outflow from rivers is 54% or 3.99×10^9 m^3/yr. of the total annual streamflow volume of 7.37×10^9 m^3/yr. available in 1993-94.

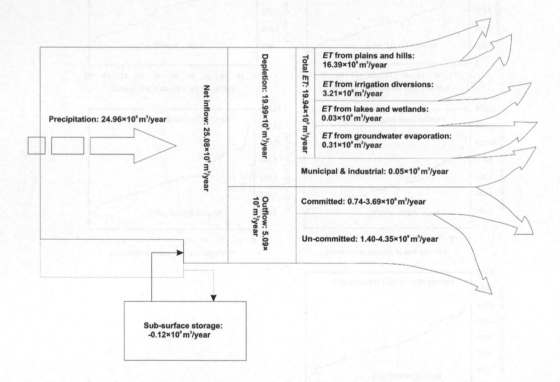

Figure 18. Finger diagram presentation of the basin level water accounts of the Karkheh river basin for the hydrological year 1993-94.

Table 8. Basin-level water accounts of the Karkheh Basin for the year 1993-94.

Water accounting indicators	Value (10⁹ m³/yr.)	Total (10⁹ m³/yr.)
	$(10^9 \text{ m}^3/\text{yr.})$	$(10^9 \text{ m}^3/\text{yr.})$
Inflow		24.96
	24.96	
Gross inflow	*0*	
Precipitation		-0.12
inflow from outside of the basin	*0*	
Storage Change	*-0.12*	
Surface		**25.08**
Sub surface		
Net Inflow		
Depletion/Consumption		19.99
Actual evapotranspiration (*ET*)	19.94	
ET from plains and hills (including all land uses)	*16.39*	
ET from Irrigation diversions to agriculture	*3.21*	
ET from lakes and wetlands	*0.030*	
ET from groundwater evaporation	*0.31*	
Municipal and Industrial	0.05	
Outflow from basin		
Total outflow		**5.09**
Surface outflow from rivers	*3.99*	
Surface outflow from drains	*1.10*	
Subsurface outflow	*0.00*	
Committed water (assumed for environment)		0.74 to 3.69*
Uncommitted outflow (Total outflow–Committed water)	5.09-3.69 to 5.09 0.74	1.40 to 4.35

Notes: Data Source: JAMAB 1999. * Values are calculated based on 10 and 50%, respectively, of the total annual streamflows (7.374×10⁹ m³/yr.) required for in-streamflows, as suggested in Tennant 1976.

Tennant (1976) suggested that 50% of the available freshwater flows are required to maintain excellent conditions in associated river ecosystems and the level of the minimum environmental flow requirements is 10%, though degradation of ecosystems will be inevitable at this level of appropriation. In many countries, the flow equivalent to *Q90* (e.g., in Brazil and Canada) or *Q95* (Australia and United Kingdom) are taken as the minimum environmental flow requirements (Tharme 2003). Based on the values suggested by Tennant (1976), we estimated committed water essentially required to support river ecosystem functions in the range of 0.74×10^9 to 3.69×10^9 m³/yr. It should be noted this is a very simple way to estimate environmental flow requirements and does not account for specific species/life phase habitat requirements, short-long-term changes in flow rates, and seasonal variability or channel geometry. Most of the environmental flow assessment studies recommend that to keep healthy, resilient and productive river ecosystems, water management policies should aim to restore the natural flow regime of the rivers, including flow variability, as much as possible (e.g., Poff et al. 1997; Richter et al. 1997). This requires detailed assessment of the flow characteristics of the Karkheh Basin streams (e.g., magnitude, timing, frequency and duration, rate of change, floods and low flows, etc.) and to explore further how to make balanced allocations to environment and human demands under varying present and future flow conditions.

However, based on these simple assumptions on uncommitted outflow from rivers, in a year like 1993-94, available for further allocation to various uses, would be in the range of 1.070×10^9 to 4.02×10^9 m^3/yr. The situation in 1993-94, when viewed in terms of future water allocation planning (Table 1), clearly highlighted the high level of competition between environmental and human demand. The water allocation to different sectors for 2001 was 4.95×10^9 m^3/yr., which is about 60% of the total renewable water resources available during the reference year 1993-94. The allocation to different sectors will be 8.90×10^9 m^3/yr. by the year 2025. Among them the irrigation share will be the biggest (7.42×10^9 m^3/yr.), which is almost equal to the renewable water supplies in an average year. The flow duration analysis suggests that planning on the basis of mean annual flows cannot provide the required streamflows every year. The anticipated situation in low flow years may be more stressful to the ecosystem health, if water allocations to human uses remain at the same levels. This also highlights the increasing stress on groundwater resources that are already overexploited, in some areas, particularly in the Gamasiab subbasin (JAMAB 2006a) and greater challenge for managing dam supplies for hydropower generation, irrigation and environment. Water allocated to the environmental sector is fixed to around 0.5×10^9 m^3/yr. (Table 1), which is even below 10% of the streamflows available in the reference year 1993-94. This indicates that further studies are required to assess the reasonable allocations for the environment, also looking into the temporal patterns of streamflows whereby streamflows should follow, to some extent, the natural patterns of flow variability. The management of releases from the newly constructed Karkheh Dam and other reservoirs would be critical to attain that, and will require more detailed scientific studies. Although the Karkheh Dam is a carryover dam, and therefore, water stored during high flow years can be used to meet demands during dry years. However, meeting the demands of all sectors would be extremely difficult in the future, particularly during dry years. Its additional complications were studied by Karamouz et al. (2006) who examined the possibilities of conflicts arising among urban, agricultural and environmental sectors located downstream of the Karkheh Dam due to deterioration of water quality as a result of increased water allocation to agriculture and urban sectors under the current water development and allocation policies. They have shown that if the current water development planning is followed, then by the year 2021 the quality of water flowing to the Hoor-Al-Azim Swamp would be deteriorated to the unacceptable levels during most of the time in a year as a result of the decreased quantity of flows and high salinity and agrochemical loads coming from agricultural return flows.

The water accounting exercise has generated useful information on the availability of water and different pathways by which water resources were depleted or moved out of the basin. The estimation of committed and uncommitted outflows provided practical insights into the degree to which water resources can be further developed. The analysis also highlighted trade-offs between different uses of water, for instance, increasing allocations to irrigation will increase the depleted portion of the water accounting and consequently reduce outflow from the basin that are likely to have a negative impact on the environment. In sum, this exercise is a simple way of viewing current pathways of water in the basin, and comparing it with variability in water supplies and future water allocations indicated trade-offs among different sectors of water use.

3.4. Concluding Remarks

This study demonstrates that the hydrology of the Karkheh Basin is governed by the natural climatic characteristics of a semi-arid to arid region, which has unique interactions with its diverse drainage areas, mostly located in the Zagros mountains. High spatio-temporal variability is a strong feature of the hydrology of the Karkheh Basin. For instance, the variability of streamflows within a month and between the months is quite high, as indicated by the range of CV values ranging from 0.44 to 1.77. The highest variability is found in November whereas the lowest variability is associated with February. In spatial terms, the highest variability is observed for Pole Chehre and Ghore Baghestan, both located in the upper parts of the Karkheh Basin.

The flow duration analysis presented in this thesis has generated further insights into the hydrological variability, surface water availability and its expected water security levels. For instance, the analysis has clearly shown that the flow regime of Pole Chehre and Ghore Baghestan (i.e., upper parts of the basin) is dominated by quick flow, whereas, base flow contributions are higher for Pole Dokhtar (i.e., middle parts of the basin) indicating a stabler flow regime for the latter station. The FDC analysis at the annual scale further reveals that the mean annual surface water availability has a security level of about 45%, ranging from 40 to 52% at the studied gauging stations across the Karkheh Basin. For example, the mean and median surface water availability at the Paye Pole station at the Karkheh River was estimated as $5,827 \times 10^6$ m^3/yr. and $5,590 \times 10^6$ m^3/yr. Like all other stations, the minimum and maximum had a wide range at Paye Pole, with values of $1,916 \times 10^6$ m^3/yr. observed during 1999-2000 and $12,596 \times 10^6$ m^3/yr. observed during 1968-69. The FDC analysis reveals that the amount of surface water available for 30 out of 40 years over the period 1961-2001 (e.g., 75th percentile, $Q75$) at Paye Pole was $4,082 \times 10^6$ m^3/yr. Furthermore, the FDC analysis has generated information on the values of various exceeding percentiles of streamflows, which could serve as the basis for water allocation planning.

The examination of water availability, variability, water accounting, and allocation planning suggested that, on the whole, water allocations to different sectors were lower than the totally available resources and, hence, the competition among different sectors of water use was not alarming during the study period. This was exemplified, for instance, by the facts that about half the total renewable streamflows was flowing out of the basin during 1993-94, the amount which is generally considered sufficient to maintain healthy ecosystems, as indicated by the records at the Hamedieh station. However, considering the high range of variability of the streamflows, changes in climate, land use and future water allocation planning, it would be extremely difficult to meet the demands in future, i.e., by 2025, as planned allocation will reach close to the annual renewable water resources available in an average climatic year. The competition between irrigation and other sectors will increase further, particularly during dry years. The analysis conducted in this study is helpful in gaining further insights into the hydrological variability of surface water resources and can, in turn, be instructive for the (re)formulation of a sustainable water resources development and management regime for the Karkheh Basin.

4.   STREAMFLOW TRENDS AND CLIMATE LINKAGES[6]

4.1.   Introduction

Examining streamflow records for the detection of trends has received increased attention of the scientific community over the last two decades (e.g., Lettenmaier et al. 1994; Zhang et al. 2001) due to the growing need to secure water for human uses such as hydropower and irrigation, as well as for aquatic ecosystems. In addition, rising concern about climate change and its impacts on streamflow has been an important driver of such studies (e.g., Cullen et al. 2002; Birsan et al. 2005).

Zhang et al. (2001), Burn and Elnur (2002) and George (2007) have illustrated that significant changes in the hydrological regime of Canadian rivers were strongly related to changes in precipitation and temperature. However, the observed trends and climate linkages were not uniformly distributed spatio-temporally. Similar observations were made by Lettenmaier et al. (1994) for the Continental United States. Cullen et al. (2002) studied the relationship of monthly streamflow, precipitation and temperature with the North Atlantic Oscillation (NAO) Index for five Middle East rivers for 1938-1984. The study indicated that changes in NAO strongly influence winter streamflows and climate. Strong linkages of NAO with temperature in the Middle East were also reported by Mann (2002). Cullen et al. (2002) stressed that, as increased greenhouse gases promote NAO's upward trend, future precipitation and winter flows will continue to decline in the study region.

Tu (2006) analyzed streamflow trends for the Meuse River Basin in Europe and suggested that the streamflows were stable at an annual time scale, though some significant increases were observed in spring flows and the flood regime. The study concluded that most of these trends were related to climate variability and were linked to changes in precipitation, which were strongly influenced by the changes in the NAO and European atmospheric circulation patterns. Ceballos-Barbancho et al. (2008) studied trends in annual and seasonal records of streamflow, precipitation and temperature for the Duero River Basin, Spain for the period 1957-2003. They found a decreasing trend in streamflow, which was strongly correlated to precipitation. They also attempted to relate changes in streamflow with land use changes (forest cover), but concluded that changes in plant cover were too far below the level of making a significant impact on the streamflow. They also noted that time-dependant

---

[6] This chapter is based on paper Streamflow trends and climate linkages in the Zagros Mountains, Iran by Masih, I., Uhlenbrook, S., Maskey, S. and Smakhtin, V. 2011. *Climatic Change* 104: 317-338. DOI: 10.1007/s10584-009-9793-x.

changes in the catchment characteristics other than climate were masked by the high inter-annual variability of precipitation in the studied Mediterranean Region. Birsan et al. (2005) studied streamflow trends in Switzerland for the period 1931-2000. Their study concluded that the mountain basins are the most vulnerable environments from the point of view of climate change. These findings are in line with those of Beniston (2003) who emphasized the importance of, and need for, more research and policy adaptation on the environmental change in the mountainous basins across the world.

Most of the climatic studies in Iran have focused on studying precipitation variability and classifying the country into different climatic regions (e.g., Domroes et al. 1998; Dinpashoh et al. 2004; Alijani et al. 2008), trend detection in the observed climatic data (e.g., Modarres and da Silva 2007) and studying the large-scale atmospheric circulations and their linkages with the local climate (e.g., Nazemosadat and Cordery 2000; Alijani 2002). To date streamflow trends and their linkages with climate are not well understood in Iran. Filling this gap is important because a) Iran is primarily an arid country with high climatic variability, b) four of its major rivers (Dez, Karun, Karkheh and Zayandeh Rud) originate from the Zagros mountains and are, thus, vulnerable to climate change with potentially adverse subsequent impacts for hydropower, agriculture and environment in the country. Therefore, exploring the fundamental questions on the nature and scale of the changes in climate and water availability is critical for informed water management and adaptation.

The main objective of this work is to identify, quantify and analyze recent trends in streamflow, precipitation and temperature using the mountainous, semi-arid Karkheh River Basin as an example. The relationship between the NAO index with the local climate (precipitation and temperature) is also investigated.

4.2.   Data and Methods

4.2.1. Hydrological and climate data and indices

For the analysis of streamflows, five stations located in the main rivers, namely Pole Chehre, Ghore Baghestan, Holilan, Kashkan and Jelogir, were selected (Figure 19 and Table 3). The daily streamflow records used in this study cover a 40-year period from October 1961 to September 2001. The months October and September refer to the start and end of the hydrological year, respectively. The stations below the Karkheh Dam were not included in this analysis because the presence of large hydrological storages and diversions are likely to obscure the relationship between streamflow and climate (e.g., George 2007). The stations used in this study generally represent the natural flow variability induced by the climatic and other physiographic factors. At some locations, water is diverted directly from streams for agricultural purposes.

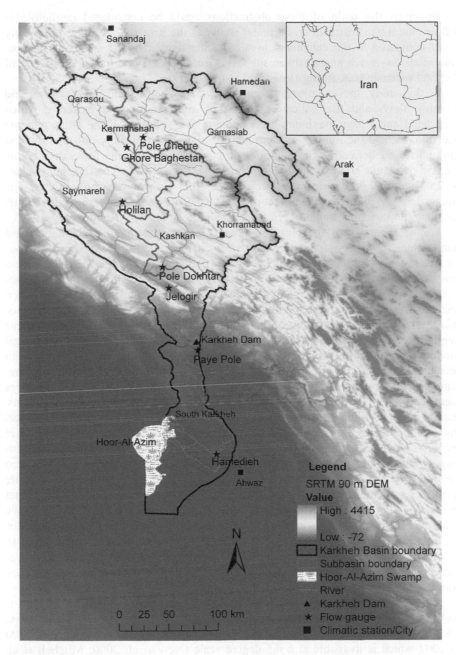

Figure 19. Location of the Karkheh Basin in Iran and some of its important features.

However, the scale of these abstractions could be considered negligible in influencing the mainstream rivers under study, as the irrigated areas are reasonably small compared to the catchment area, e.g., the total irrigated areas in Qarasou, Gamasiab, Saymareh and Kashkan subbasins were 5, 12, 3 and 6%, respectively, of the total subbasin area during 1993-94 (JAMAB 1999).

In this study, the examined streamflow variables were the mean annual and monthly flows and the indicators describing the hydrological extremes which included 1 and 7 days maximum and minimum flows, timing of the 1-day maxima and minima, and the number and duration of high- and low-flow pulses. Low and high pulses are defined as those periods during which daily mean flows drop below the 25th percentile and exceed the 75th percentile, respectively. The threshold values of 25th and 75th flow percentiles are derived from a flow duration analysis. The streamflow variables selected reflect different aspects of a natural river flow regime, i.e., magnitude, timing, duration, frequency and rate of change (Richter et al. 1997).

For the analysis of the climate, monthly climatic data on precipitation and temperature for six synoptic stations, two located in the Karkheh Basin and four located in the vicinity of the basin were used (Figure 19 and Table 9). There were a few other climatic stations located inside and close to the basin, but they were not used in this study because of shorter and incomplete records. The precipitation analysis was confined to the months of October through May, when about 99 % of the total annual precipitation occurs, and data for other months with almost negligible rainfall were not analyzed. The indicators used in the study were total monthly precipitation, number of precipitation days, number of days with precipitation equal to, or greater than, 10 mm/d, maximum daily precipitation, number of snow and sleet days and mean monthly temperature. These indicators were selected from the list of available climatic indices at the selected stations, mainly due to their importance in the hydrological processes. For instance, maximum daily precipitation is likely to influence flood response and groundwater recharge processes. Similarly, precipitation values of more than 10 mm/d are also important for runoff generation process. The monthly NAO index for the same period was obtained from the CRU website (http://www.cru.uea.ac.uk/ftpdata/nao.dat). Since the NAO index is known to impact the winter climate of the Middle East (e.g., Cullen et al. 2002; Mann 2002), we examined its relationships with the local climate (precipitation and temperature) for the four winter months of December to March in the study area. In addition to monthly correlations, the relationship of the composite NAO index, averaged over December-March, with the corresponding values of precipitation and temperature was also investigated.

Although the studied station covered about 50 years, from the 1950s to 2003, it is worth looking at climatic patterns spanning over the last century. For this purpose, CRU data on monthly precipitation and temperature were used for the period 1900 to 2002 which is available at a 0.5 degree scale (New et al. 2000; Mitchell et al. 2004) available through http://www.cru.uea.ac.uk/~timm/grid/CRU_TS_2_1.html.

Streamflow Trends and Climate Linkages

Table 9. Geographic and climatic characteristics of the selected synoptic climatic stations.

Name of station	Longitude (degrees East)	Latitude (degrees North)	Elevation (masl)	Record length, temperature	Record length, precipitation	Annual temperature (°C)		Annual evaporation demand (mm/yr.)		Annual precipitation (mm/yr.)	
						Mean	Standard deviation	Mean	Standard deviation	Mean	Standard deviation
Kermanshah	47.70	34.17	1,322.0	1951-2003	1951-2003	14.2	0.8	1,515	58	447	134
Sanandaj	47.00	35.20	1,373.4	1960-2003	1960-2003	13.4	0.9	1,437	60	464	122
Hamedan Nozheh	48.43	35.12	1,679.7	1955-2003	1951-2003	11.0	1.0	1,362	58	335	87
Khorramabad	48.22	33.29	1,125.0	1951-2003	1951-2003	17.2	1.1	1,602	56	510	128
Arak	49.46	34.6	1,708.0	1955-2003	1955-2003	13.8	1.1	1,364	57	342	102
Ahwaz	48.40	31.20	22.5	1957-2003	1957-2003	25.3	0.8	1,930	57	229	87

Notes: Potential evapotranspiration was calculated using the Hargreaves method (Hargreaves et al. 1985). Mean and standard deviations for temperature, evaporation and precipitation are based on the annual values.
Data source: Islamic Republic Iran, Meteorological organization (http://www.irimo.ir/english/).

The records of the available stations had very few missing values. The missing values were replaced by average values in case of climatic data. The missing values in daily streamflow were filled by taking average of the days before and after the day having missing record. The missing records for longer periods were filled by using statistical relationship based on correlation analysis with the neighboring station(s).

4.2.2. Trend and correlation analysis

Trends were examined using the Spearman's Rank (SR) test (e.g., McCuen 2003; Yue et al. 2002). The studied data were examined for the presence of serial correlation before conducting the trend analysis. The serial correlation, if found significant, was removed using the pre-whitening method, before application of the trend test. The relationship between streamflow and climatic variables was studied by performing a correlation analysis among them. This analysis was mainly focused on the two catchments, Qarasou and Kashkan, mainly because of the better representative climatic data sets. The climatic data of Kermanshah and Khorramabad were used to study the linkages between streamflows at Ghore Baghestan and Pole Dokhtar, respectively.

4.3.   Results and Discussion

4.3.1. Characterizing the streamflow regime

A brief description of the salient features of the streamflow regime, in terms of study variables, is presented in this section. Table 10 shows the mean and CV (given in parenthesis) of the studied streamflow variables. The results substantiate that the studied streamflow variables generally have high variability. For instance, the differences between the peak and low flows within a year are quite large. For example, at Jelogir, the mean monthly streamflow in April (386 m^3/s) is nearly ten times higher than in September (41 m^3/s). Although the extreme floods (i.e., 1-day maximum) are generally observed in spring, particularly in March, these can occur any time from November to April (Figure 20). On an average, 2 to 5 high pulses are observed in a year, with the mean duration ranging from 8 to 17 days. The peaks are generated as a result of the large amount of precipitation as well as from snowmelt contributions or the combined effect of snowmelt and rainfall. The low flows are recorded from June to September. During this period the magnitudes are quite low, though there remains some water in all the main rivers throughout the year. On an average, 2 low pulses in a hydrological year, with a mean duration ranging from 63 to 79 days is observed across the examined stations.

Table 10. Streamflow characteristics indicating mean and CV (given in parenthesis) values at selected locations in the Karkheh River Basin.

Streamflow indicators	Units	Pole Chahre	Ghore Baghestan	Holilan	Pole Dokhtar	Jelogir
Annual	m³/s	34(0.50)	23(0.54)	77(0.52)	52(0.41)	158(0.42)
October	m³/s	7(0.76)	5(0.47)	16(0.53)	20(0.45)	55(0.40)
November	m³/s	25(1.77)	12(1.26)	46(1.42)	35(0.94)	108(1.18)
December	m³/s	32(0.76)	16(0.68)	62(0.69)	47(0.61)	143(0.60)
January	m³/s	34(0.54)	20(0.62)	72(0.55)	51(0.54)	157(0.53)
February	m³/s	46(0.44)	30(0.53)	105(0.46)	74(0.52)	221(0.43)
March	m³/s	95(0.70)	65(0.86)	222(0.89)	123(0.52)	379(0.57)
April	m³/s	98(0.63)	64(0.69)	220(0.71)	123(0.59)	386(0.59)
May	m³/s	52(0.81)	36(0.66)	117(0.74)	74(0.74)	230(0.65)
June	m³/s	11(0.84)	13(0.66)	32(0.74)	28(0.55)	85(0.57)
July	m³/s	5(0.85)	6(0.70)	16(0.77)	19(0.46)	52(0.52)
August	m³/s	3(0.87)	4(0.67)	10(0.66)	16(0.39)	41(0.45)
September	m³/s	3(0.85)	4(0.58)	10(0.60)	15(0.35)	39(0.39)
1-day minimum	m³/s	2(0.71)	2(0.58)	6(0.56)	12(0.34)	28(0.44)
7-day minimum	m³/s	2(0.67)	2(0.53)	7(0.48)	12(0.32)	31(0.39)
1-day maximum	m³/s	274(0.73)	183(1.06)	613(0.91)	553(0.63)	1,093(0.64)
7-day maximum	m³/s	197(0.69)	135(0.95)	446(0.91)	265(0.52)	751(0.60)
Date of minimum	Julian day[a]	251(0.07)	254(0.09)	248(0.09)	250(0.09)	248(0.12)
Date of maximum	Julian day	82(0.10)	87(0.08)	103(0.17)	68(0.13)	71(0.13)
Low pulse count	No.	2(0.83)	2(0.72)	2(0.75)	2(1.14)	2(1.18)
Low pulse duration	Day	63(0.78)	68(0.76)	79(0.67)	71(0.86)	68(0.77)
High pulse count	No.	3(0.70)	2(0.68)	3(0.91)	5(0.58)	4(0.61)
High pulse duration	Day	13(0.94)	15(0.98)	17(1.09)	8(1.05)	11(0.88)

Note: [a]Julian day is calculated for a calendar year based on the notion of taking 1st January as 1st Julian day and 31st of December as 365th or 366th Julian day.

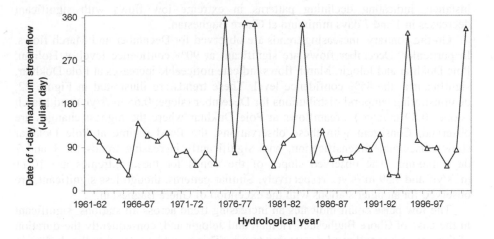

Figure 20. Timing of the 1-day maximum streamflow, illustrated by the records at Pole Dokhtar.

4.3.2. Streamflow trends

The results of the trend analysis for the streamflow variables are presented in Table 11. The sign of the t statistics indicates the nature of the trend, with positive and negative values indicating increase and decrease, respectively, and its magnitude indicates the level of significance. Trends are not significant for the mean annual streamflows across any of the examined locations. However, a number of significant trends (both increasing and decreasing) have been found with respect to other indicators. This stresses the need for investigating streamflow trends at finer temporal resolutions, as most of the important trends may not be identified at the coarser time scales.

Streamflows in May depict a decreasing trend for all examined stations, though more noticeable in upper parts of the basin. The trend is observed significant at Ghore Baghestan and Holilan as shown in Figure 21. The slope of the linear trend indicated a decrease of about 0.61 m³/s/yr. at Ghore Baghestan and 1.94 m³/s/yr. at Holilan. A strong decline at Holilan is due the compound effect of the upper two stations, both experiencing a decline. The visual inspection of the inter-annual patterns from Figure 21 suggests that increases and decreases follow a pattern of 2 to 5 years, with a more significant decline after the late 1970s.

However, the years with high and low flows follow similar patterns for both locations, e.g., highest and lowest values were observed during 1969 and 2000, respectively. Further to the decline in May flows at these two locations, all the low flow periods from June to September indicate a declining pattern with significant trends in August. It is noteworthy that this pattern is similar, though with varying significance, for four out of five observed stations, with the exception of Pole Dokhtar. The impact of these patterns is clearly seen in the three upper stations, for instance, indicating declining patterns in extreme low flows with significant decreases in 1 and 7 days minimum at Ghore Baghestan.

On the contrary, increasing trends are observed for December and March flows. In particular, December flows are significant at 90% confidence level at Holilan, Pole Dokhtar and Jelogir. March flows indicate noticeable increases at Pole Dokhtar, significant at the 85% confidence level. These trends are illustrated in Figure 22, demonstrating temporal distributions for December (slope: 0.65 m³/s/yr.) and March (slope: 0.89 m³/s/yr.) streamflows at Pole Dokhtar where the highest changes are observed. Consistent with these observations, the flood regime at Pole Dokhtar shows discernible intensification, with significantly increasing trends for 1 and 7 days maxima. The observed slopes of the trends for these extremes are 10.11 m³/s/yr. and 3.59 m³/s/yr., respectively. Similar patterns, though less significant, are observed for all other stations with the exception of Ghore Baghestan.

The low pulse count indicates an increasing trend across all stations, significant in the case of Ghore Baghestan, Holilan and Jelogir and, consequently the duration of the low pulses followed decreasing trends. This could be related to the decline in the low-flow regimes because of their interrelationships. Similarly, increasing patterns of high pulse duration and high pulse count are generally consistent with the increasing trends in the flood regime. Therefore, recognition of the interdependence of the studied streamflow variables is helpful while interpreting the consistency of the trend results. For example, the 7-days maximum flow is strongly correlated with 1-day maximum, and the 7-days minimum flow follows pattern similar to that of 1-day minimum. Moreover, these maximum and minimum flows are governed by the variability in the corresponding streamflows during high- and low-flow periods, which also have an influence on the dynamics of associated high- and low-flow pulses.

Table 11. Results of the trend analysis showing calculated t statistics for streamflow indicators.

Streamflow variables	Pole Chehre	Ghore Baghestan	Holilan	Pole Dokhtar	Jelogir
Annual	0.479	0.0860	0.180	1.140	0.680
October	-0.594	0.471	-0.075	1.467*	0.689
November	-0.575	0.153	0.204	0.732	0.271
December	0.460	0.840	1.315*	2.070*	1.385*
January	*-0.212*	0.329	0.690	0.993	0.325
February	-0.459	-0.354	0.104	*0.463*	0.315
March	0.671	0.328	0.796	1.248	1.152
April	0.345	-0.049	0.068	0.631	0.522
May	-1.193	-1.535*	-1.543*	-0.014	-0.806
June	-0.808	-1.119	-1.251	0.278	-0.408
July	-0.987	-1.129	-1.160	0.428	-0.314
August	-0.837	-1.446*	-1.461*	0.293	-0.174
September	-0.590	-1.020	-0.911	0.579	0.045
1-day minimum	-0.578	*-1.763**	*-1.01*	*-0.378*	*-0.625*
7-day minimum	-0.132	*-1.560**	*-0.624*	*-0.069*	*-0.238*
1-day maximum	1.269	-0.234	0.081	1.996*	1.347*
7-day maximum	1.063	0.163	0.200	2.298*	1.052
Date of minimum	0.382	-0.566	-0.534	0.156	-0.548
Date of maximum	-1.282	-0.012	-0.826	-0.196	-0.085
Low pulse count	1.085	3.307*	2.991*	0.843	2.368*
Low pulse duration	0.227	-1.446*	-0.906	-1.469*	-1.722*
High pulse count	0.332	-0.428	0.178	1.670*	1.293
High pulse duration	1.791*	1.185	0.881	0.930	0.636

Notes:
[a] Numbers in italics refer to the trend results based on the pre-whitened data.
* Indicate a significant trend at 90% confidence level (one tailed).

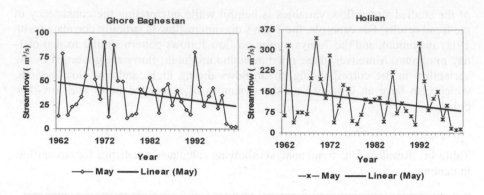

Figure 21. Declining trend in May streamflow at Ghore Baghestan and Holilan.

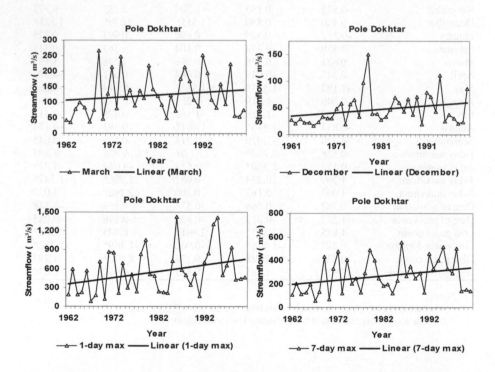

Figure 22. Increasing trends observed at Pole Dokhtar, illustrated by December, March, 1 and 7 days maximum streamflows.

It is important to note that the serial correlation was mainly depicted by the extreme low flow indicators (e.g., 1 and 7 days minimum flows at four out of five examined stations indicated significant r_1). The rest of the studied indicators did not depict persistence in them, with the exception of January flows at Pole Chehre and February flows at Pole Dokhtar. The strong persistence in the extreme low flow indicators could be due to the groundwater flow processes which are slow in nature and, consequently, manifest a carry over effect over time. This observation of persistence in the low flow indicators is in agreement with the literature (e.g., Douglas et al. 2000). Moreover, the presence of serial correlation did not alter the trend results as the number of significant trends remains the same with and without the pre-whitening. Although the magnitudes of the calculated t statistics were different in the case of the pre-whitened series when compared to their corresponding values without pre-whitening, these were not distinctive enough to change the indicated significance level. For example, t values for 7-days minimum flow at Ghore Baghestan were estimated as 1.560 and 1.670 in the cases of with and without pre-whitening, respectively. Both of these values constitute a significant category at 90% confidence level. Moreover, insignificant impact of pre-whitening could be due to the lower values of the first serial correlation (r_1) observed in this study, which falls around 0.3 in most cases of the pre-whitened streamflow indicators. This point is substantiated by the study of von Storch and Navarra (1995). They demonstrated that the false rejection of the null hypothesis also depends on the magnitude of r_1. They showed that while applying the Mann-Kendall trend test the chances of false rejection of the null hypothesis increase from 15% at r_1 value of about 0.3 to more than 30% when r_1 exceeds 0.6.

4.3.3. Trends in the climatic data

The trend investigation results for precipitation and temperature are given in Table 12. As in the streamflow indicators, the number of significant trends remains the same before and after the pre-whitening of some of the studied climatic data series, most likely due to the reasons similar to those mentioned before. The results indicate that the total annual precipitation has not significantly changed at most of the examined stations, with the exception of Arak where a downward trend is significant. However, a number of upward and downward trends are observed in other studied indicators at an annual scale (Table 12). These trends were quite consistent with the changes observed in the studied indicators at the monthly scale at the respective stations. On the whole, reasonably uniform trends in precipitation are observed on various indicators for precipitation in April, May, March, October and December precipitation. Among them, the decreasing trend in total monthly precipitation in April is the most striking, which is significant for the four out of six stations. Similarly, a decreasing pattern is observed for May, though less significant compared to that for April. The results indicate that the precipitation regime shows a general decreasing trend during these 2 months for the Karkheh Basin as well as for the neighboring areas in the Zagros mountains.

Table 12. Trend results for precipitation (P) and temperature (T) data showing calculated t statistics for the studied climatic indicators.

Variable/ Climatic station	Oct.	Nov.	Dec.	Jan.	Feb.	Mar.	Apr.	May	Annual
Total P									
Kermanshah	1.247	-0.044	1.229	1.476*	0.129	0.947	-2.458*	-0.618	0.788
Sanandaj	0.302	0.025	0.688	-1.839*	-1.144	-0.572	*-0.548*	-1.203	*-1.242*
Hamedan	1.463*	-0.376	0.685	0.292	-0.745	0.164	*-1.615**	-0.939	-1.239
Khorramabad	0.799	-0.933	0.531	-0.752	0.341	1.279	-1.379*	-0.637	-0.613
Arak	1.238	-0.076	0.529	-3.122*	-1.61*	0.39	*-1.368**	-0.061	-1.972*
Ahwaz	0.565	1.010	0.896	0.744	0.320	2.551*	-0.728	-0.973	1.221
No of P days									
Kermanshah	1.636*	0.509	2.126*	2.586*	0.652	0.465	-1.802*	0.708	1.813*
Sanandaj	0.79	0.816	0.951	-0.781	-0.888	-0.598	-2.257*	-0.918	-1.048
Hamedan	1.503*	0.252	1.778*	1.595*	0.728	-0.063	-0.666	0.281	1.506*
Khorramabad	1.495*	-0.491	0.7	1.285	0.171	0.76	-0.738	-0.042	0.557
Arak	1.185	1.102	2.016*	0.745	1.131	*1.518**	-0.388	0.288	1.934*
Ahwaz	1.769*	1.548*	3.589*	3.244*	3.555*	*3.046**	1.673*	1.657*	*4.572**
No of days P ≥10 mm									
Kermanshah	1.733*	-0.46	0.611	1.356*	0.359	0.753	-1.95	-0.706	0.254
Sanandaj	0.81	*-0.01*	0.884	-1.431*	-0.203	0.207	-0.321	-0.933	*-0.672*
Hamedan	1.995*	0.122	1.236	0.393	-0.761	1.253	*-1.258**	-0.728	-1.09
Khormabad	1.371*	0.028	0.278	-0.289	1.518*	1.348*	-0.63	-0.181	-0.227
Arak	0.831	-0.079	0.569	-2.015*	-1.821*	0.217	*-1.42**	0.354	-2.373*
Ahwaz	1.352*	1.902*	0.966	0.479	0.951	2.133*	0.518	0.714	1.543*
No of snow/sleet days									
Kermanshah			-0.336	1.483*	0.742	-0.529			0.319
Sanandaj			0.111	-0.594	0.178	-1.223			-0.451
Hamedan			0.676	1.824	1.019	-0.025			1.109
Khormabad			0.793	0.69	0.455	1.216			0.264
Arak			1.399*	0.484	1.524*	1.189			1.259
Greatest daily P									
Kermanshah	1.158	-0.052	1.709*	1.062	0.281	1.576*	-1.117	-0.818	-0.273
Sanandaj	0.305	-0.401	0.354	*-1.472**	-2.180*	-0.608	-0.541	-1.051	-0.936
Hamedan	1.433*	-0.452	0.641	-0.767	0.081	0.017	-2.845*	-0.689	-1.582*
Khorramabad	0.800	-0.609	0.370	-1.093	0.193	2.735*	0.900	-0.780	0.85
Arak	1.277	0.192	0.420	-3.055*	-1.250	0.345	-0.388	0.619	-2.042*
Ahwaz	0.757	0.992	0.841	1.172	0.054	2.378*	-0.950	-1.063	0.746
Mean T									
Kermanshah	*2.586**	3.531*	2.266*	0.750	0.935	1.389*	3.896*	4.003*	*3.904**
Sanandaj	*0.597*	0.309	1.008	0.875	0.764	0.249	3.260*	0.337	*1.421**
Hamedan	*-0.365*	-0.030	-0.558	-0.897	-1.149	-1.574*	1.744*	-0.403	*-0.175*
Khorramabad	*-2.52**	*-2.62**	-1.726*	-2.065*	-2.020*	*-2.193**	-1.752*	*-2.312**	-2.279*
Arak	*-0.688*	-0.899	-0.247	0.127	-0.304	-1.804*	0.829	-0.945	-0.427
Ahwaz	2.830*	2.101*	1.853*	0.826	1.197	0.760	3.070*	4.055*	*3.382**

Notes:
[a] Numbers in italics refer to the trend results based on the pre-whitened data.
* indicates a significant trend at the 90% confidence level (one tailed).

Conversely, precipitation in March shows an increasing trend in most cases, except for Sanandaj. For the Karkheh Basin, the trend is noteworthy for Khorramabad where significant trends are observed for total precipitation (significant at 85% confidence level), number of days with precipitation ≥ 10 mm/d (significant at 90% confidence level) and the amount of greatest daily precipitation (significant at 90% confidence level). December precipitation also exhibits an increasing trend in terms of monthly totals, with the number of days with precipitation showing an increasing trend at most of the stations. Precipitation totals in October show an increasing pattern, with significant trends for Hamedan. Mean monthly temperature showed nonuniform trends across the examined locations, and most noteworthy for the study region are the increasing trends at Kermanshah and decreasing trends at Khorramabad.

Some of the abovementioned trends regarding precipitation show similarity with an earlier study conducted for semi-arid to arid regions of Iran by Modarres and da Silva (2007). They studied trends in total annual precipitation, total monthly precipitation and total number of rainy days in a year. They used data of 20 climatic stations located all across Iran with varying recorded lengths of time of 32 to 50 years. It is important to note that their study period was almost similar to ours but the subset of climatic stations was different from those used in our study and none of their stations fall within the Karkheh Basin and its proximity. Nonetheless, their findings are largely consistent with those of our study. Their results show no significant trends in total annual precipitation and number of rainy days in a year at 18 out of 20 stations. They reported the presence of increasing and decreasing trends in spring and winter months across a few locations, most notably a significant decrease in precipitation during April at four stations and significant increases at four stations in March.

4.3.4 Streamflow trends and climate linkages

The correlation analysis indicates that the mean temperature showed negative correlation with streamflow variables, but these correlations were generally weak by themselves (less than 0.3 in most cases) and also in comparison with those of precipitation with streamflow. Therefore, further discussion is mainly focused on streamflow and precipitation. Streamflows are strongly correlated with precipitation at the annual scale as indicated by the correlation analysis between average annual streamflow at Ghore Baghestan and total annual precipitation at Kermanshah ($r = 0.81$). A similar inference was drawn from the correlation analysis between streamflow at Pole Dokhtar and precipitation at Khorramabad ($r = 0.84$). Furthermore, both variables did not exhibit significant trends at annual scale at the studied locations (Tables 11 and 12). The presence of a strong correlation at annual scale is in agreement with an earlier study in the Karkheh Basin by Sutcliffe and Carpenter (1968).

However, one can anticipate that the relationship of other streamflow variables, e.g., monthly flows and extremes, with precipitation may not be straightforward because of the complexities of various hydrological processes. For instance, mean flow in a month may not be entirely dependent on the precipitation in that month, but is often the result of a combined effect of precipitation during that and earlier months. For the two selected subbasins of the study area (e.g., Qarasou subbasin represented by the streamflow measured at Ghore Baghestan station and Kashkan subbasin represented by the streamflow measured at Pole Dokhtar station), the correlation analysis with monthly streamflows and precipitation suggests that the precipitation in a month can influence monthly streamflows with a lag time varying from 1 to 7 months. The main causes are the occurrence of snowfall in winter which mainly melts in spring and the contribution of (delayed) runoff from the subsurface storage. For instance, at Ghore Baghestan, streamflows in November are influenced by precipitation in November (r = 0.79) and October (r = 0.38), whereas streamflows in June are governed by the precipitation in January to May (r ranging from 0.32 to 0.57). The presence of a good correlation among most of the streamflow and precipitation variables enabled us to develop the linkages between them, which are discussed in detail further below. We also conducted a correlation analysis by using the Spearman and Kendall methods (e.g., McCuen 2003), not shown here because the results were in good agreement with those of the Pearson method.

The Case of Ghore Baghestan

The decline in May streamflow could be attributed to the decline in precipitation during April and May. Both these months strongly influence the streamflow in May, as indicated by the strong correlation between streamflows and precipitation (r = 0.56 for April and r = 0.49 for May). The analysis of the groundwater contribution to the total streamflow during May could further help understand the dependency of May streamflow on the precipitation in the previous months. Masih et al. (2009) have estimated a base flow index of 0.7 for May at Ghore Baghestan using long-term data of the period 1961-2001. This clearly depicts a very high contribution from groundwater storage in the low flow regime. Considering no significant change in precipitation during winter periods (December to March) and a noticeable decline in May and April precipitation in the area, it can be argued that these changes in the precipitation are the main trigger of declining streamflows in May. This point is further substantiated in Figure 23, which clearly shows the streamflow in May following similar patterns of precipitation in April and May. The CRU data helped to look at trends over the last century (1901 to 2002). Further examination of CRU data produced similar evidence that the precipitations during April and May have significantly decreased over time, notably, after the 1980s.

Since the low flow regime of the Ghore Baghestan is strongly governed by the precipitation in April (r ranging from 0.54 for June to 0.44 for September) and May (r ranging from 0.36 for June to 0.23 for September), the decline in monthly flows (June through September) and extreme low flow conditions (1 and 7-days minima) could also be attributed to the decline in precipitation during April and May. But, there might be some other complementary factors as well. For instance, the increasing trend in temperature observed for Kermanshah might be another reason for the decline in streamflow during low flow periods observed at Ghore Baghestan. This is likely to impact the snowfall and snowmelt dynamics in the area and could induce early snowmelt causing an increase in winter flows and a subsequent decrease in spring and early summer flows (e.g., Arnell 1999; Bouraoui et al. 2004). Although these observations were not sufficiently supported by the observed trends in the streamflow at Ghore Baghestan, they do signal towards these changes as depicted by positive t values for December, though not significant at this station (Table 11). Nonetheless, the observed upward trend in December flow at Holilan is significant ($t = 1.315$) which indicates the combined effect of the abovementioned changes observed at Ghore Baghestan and Pole Chehre. The rise in temperature is likely to accelerate ET processes and hence could reduce the streamflow (e.g., Nash and Gleick 1991; Ficklin et al. 2009). Additionally, the increase in crop water demand is likely to enhance water consumptions (e.g., more irrigation applications by the farmers). Consequently, this is likely accelerate irrigation withdrawals from the streams, which would be significant during low flow periods when streamflows are already at their lowest level (e.g., about 5 m^3/s from July to September at Ghore Baghestan, Table 10).

Generally, the trends observed at Ghore Baghestan are consistent with those observed at Pole Chehre, most notably in terms of the low flow indicators (Table 11). The composite impact of these changes observed at Ghore Baghestan and Pole Chehre is also clearly evident from the observations at Holilan station on the Saymareh River, which is mainly sourced through these two upper subbasins. However, the observed changes in these three stations were not uniform across all other examined stations. For instance, changes in the low flow regime were not significant at the Pole Dokhtar. On the other hand, the flood regime showed significant upward trends in the middle parts of the basin, which are further discussed in the following section.

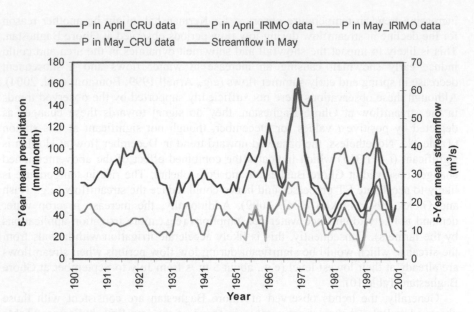

Figure 23. The linkages of trends in streamflow in May and precipitation in April and May, illustrated by the case of Ghore Baghestan.

The Case of Pole Dokhtar

The increasing trends observed in the streamflows at Pole Dokhtar are consistent with the climatic alterations observed in the area. This is evident by the increasing trends in precipitation during winter and decreasing trends in temperature observed at Khorramabad (Table 12).

The flood regime at this location, i.e., 1 and 7-days maxima mainly depends on the winter precipitation, with major influence of precipitation in March and February, as indicated in Figure 24. Therefore, intensification of the precipitation regime in March will possibly be the main cause of the increasing trends in the flood events as most of them occur in March, as indicated by Figure 20. This point is further supported by concurrent trends in the March streamflow ($t = 1.248$, significant at 85% confidence level) and total precipitation in March ($t = 1.279$, significant at 85% confidence level).

The increasing trends in December streamflows could be mainly linked to the increasing trends in the precipitation regime in December ($r = 0.67$). The streamflow in October is significantly correlated with precipitation in March ($r = 0.44$), February ($r = 0.41$) and October ($r = 0.38$). This suggests that the increasing trends in October streamflow are due to the increases in the precipitation regime in these months, most notably March and October. It can be further explained from the point of view of a hydrological process, as it is very likely that more frequent precipitation events and of greater magnitude produce more surface runoff as well as more recharge to the subsurface flows. The increased subsurface recharge contributes to the streamflows in the latter part of the year via base flow. This means that the precipitation recharging subsurface flow in March and February has an influence on October streamflows in this case. Since temperature data showed negative correlations with the streamflows, the significant decline in temperature observed for this region are likely to reduce ET and, therefore, might be another contributing factor towards the general increase in streamflows at Pole Dokhtar. Another factor contributing to these increasing trends could be the watershed degradation that has taken place in the study area over the last few decades (Ghafouri et al. 2007). Some studies suggest that the decrease in forest cover increases the flood potential (e.g., Guo et al. 2008). For the Karkheh Basin, this point is further supported by the study of Mirqasemi and Pauw (2007) that compared the land use maps derived from Landsat data for the years 1975 and 2002, and found that a decline of about 25% has occurred in the forest cover during this period in the Karkheh Basin. Therefore, this change is likely to cause increasing trends in the flood regime, particularly in the middle parts of the basin where forests are a major land cover, but this warrant further research.

The changes in the low flow regime (e.g., indicated by flows from May through September and 1 and 7 days minimum) at Pole Dokhtar were not significant, despite declining trends in precipitation in April and May for the region indicated by the climatic station at Khorramabad. This could be due to the counter-effect of the increase in precipitation during March. Another reason could be the decrease in evaporation demand, caused by the decreasing trend in temperature observed in the region, which is likely to contribute positively to the streamflow generation processes.

As expected, the observed changes in the streamflows at Jelogir station on the Karkheh River generally concur with those at the upstream stations (e.g., Pole Dokhtar and Holilan). This is particularly evident by the significant trends in December flow, 1-day maximum flow, low pulse count, and low pulse duration observed at Jelogir (Table 11).

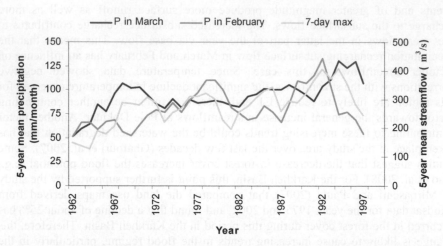

Figure 24. Linkages of extreme floods with precipitation, indicated by the 7-days maximum streamflows at Pole Dokhtar and precipitation in March and February at Khorramabad.

4.3.5   The impact of NAO index on the local climate

Further examination of the observed trends in relation to the changes in the global circulation patterns generated useful insights into the study region. The earlier studies in the Middle East have shown the influence of NAO on controlling the temperature and precipitation regime during winter and early spring (e.g., Cullen et al. 2002; Mann 2002; Zangvil et al. 2003; Evans et al. 2004). We also attempted to study the correlation between monthly NAO index with the monthly precipitation and temperature during the winter months from December to March. These relationships were also investigated for the whole winter period by averaging the data sets from December through March. The results indicated very weak correlations with precipitation (Table 13). Nonetheless, the NAO index showed comparatively better correlations with temperature (Table 13), with all the stations located in the Zagros mountain area depicting significant correlations for the composite NAO index from December to March as well as for most of the winter months.

Our findings regarding precipitation are different from those of Cullent et al. (2002) who found a strong impact of changes in the NAO on the streamflow, precipitation and temperature during December-March for the neighboring Euphrates-Tigris River system. However, our findings correspond with those of Evans et al. (2004) who found that NAO index alone could not be a predictor of the local climate in the Zagros mountains, Iran. They used climate models to simulate the climate of the Middle East, including the Zagros mountain ranges, Iran, and illustrated that local factors related to storm tracks, topography, and atmospheric stability have a strong control over climate of the Zagros mountains as compared to NAO. A study by Alijani (2002) on the linkages of 500 hpa (hectopascals) flow patterns and the climate of Iran also indicated the importance of the local climatic factors. Therefore, more detailed studies are required on linkages between long-term changes in the local climate (i.e., precipitation and temperature) and global as well as local circulation patterns.

Table 13. Correlation (r) of NAO index with winter precipitation and temperature for the study area.

Variable/Climatic station	December	January	February	March	Composite (December to March)
Precipitation					
Kermanshah	-0.209	-0.062	0.040	-0.161	-0.031
Sanandaj	-0.213	-0.403*	-0.070	-0.139	-0.247
Hamedan	0.028	-0.013	0.108	-0.049	-0.002
Khorramabad	-0.151	-0.033	0.031	-0.062	-0.090
Arak	-0.098	0.001	-0.018	-0.025	-0.277
Ahwaz	-0.039	-0.044	0.053	0.008	-0.044
Temperature					
Kermanshah	-0.322*	-0.301*	-0.350*	-0.322*	-0.344*
Sanandaj	-0.315*	-0.341*	-0.238	-0.403*	-0.291*
Hamedan	-0.407*	-0.294*	-0.274	-0.396*	-0.312*
Khorramabad	-0.302*	-0.376*	-0.432*	-0.411*	-0.484*
Arak	-0.346*	0.337*	-0.249	-0.363*	-0.346*
Ahwaz	-0.252	-0.385*	-0.223	-0.460*	-0.297

Note: * indicates significant correlation at 95% confidence level

4.4. Concluding Remarks

The study provided an overview of the changes in the streamflows in the Karkheh Basin and identified a number of trends, both increasing and decreasing. Most of these trends were found triggered by climatic factors - mainly by changes in precipitation. The most notable trends were declines in May streamflows, which can be attributed to the decline in precipitation in April and May. The two upstream catchments displayed declining trends in low flow regimes, demonstrated by monthly streamflows, 1 and 7 days minima and the number and duration of low flow pulses. In the middle part of the basin (at Pole Dokhtar) the increasing trends were reflected by 1 and 7 days maxima, and March, December and October flows. These trends can be attributed to the intensification of the precipitation regime in these months, with the March precipitation having the highest influence on the flood regime.

The observed trends for Holilan at the Saymareh River and Jelogir at the Karkheh River were a reflection of the combined effects of their upstream catchments. Similar to the observed patterns at Ghore Baghestan and Pole Chehre, the Holilan indicated declining patterns in monthly streamflows from May through September, as well as a decline in the low flow regime.

All trends were not reflected in the flow regime of the Karkheh River because of the varying changes in the upper and middle parts of the basin. The changes at Jelogir were significant when the patterns were similar for most of the upstream tributaries. For instance, consistent with the upper parts of the basin, the declining patterns from May through August were observed at Jelogir, but were not as significant as in the case of the upper stations. This is because the more stable response of Pole Dokhtar during these months counterweighted these declining patterns were observed in the upper parts of the basin. Nevertheless, the significant trends in streamflows at Jelogir, i.e., an increase in the 1-day maximum, December flows, low pulse count and a decrease in low pulse duration, indicated alterations of the hydrological regime of the Karkheh River due to the changes in climate during the study period.

Contrary to expectations, North Atlantic Oscillation Index did not show a good correlation with the precipitation in the Zagros mountains because its impact might be masked by the strong topographic controls and other local climatic factors, which deserve further research.

Since most of the observed changes in streamflow, precipitation and temperature were not uniformly distributed across the Karkheh Basin, the adaptation response should be different for different parts of the basin. If the observed trends will persist, the major policy concerns about water management would be how to a) stabilize declining streamflows during low flow periods in the upper parts of the basin, and b) control intensification of the flood regime in the middle parts of the basin.

5. REGIONALIZATION OF A CONCEPTUAL RAINFALL-RUNOFF MODEL BASED ON SIMILARITY OF THE FLOW DURATION CURVE[7]

5.1. Introduction

5.1.1. Problem statement

Streamflow data are a prerequisite for planning and management of water resources such as the design of dams and hydropower plants, assessment of water availability for irrigation and other water uses, assessment of flood and drought risks and ecological health of a river system. However, in many cases, observed streamflow data are not available or are insufficient in terms of quality and quantity. This undermines the informed planning and management of water resources at a specific site as well as at the river-basin scale.

Hydrologists have responded to this challenge by developing various predictive tools, which are commonly referred to as regionalization methods (e.g., Blöschl and Sivapalan 1995; Sivapalan et al. 2003; Yadav et al. 2007). These methods can be broadly classified into two groups based on their temporal dimension. The first group deals with the estimation of continuous time series of streamflows (e.g., Magette et al. 1976; Merz and Blöschl 2004). The second group estimates selected hydrological indices, such as the mean annual flow and base flow index (e.g., Nathan and McMahon 1990b), or various percentiles of the flow instead of continuous time series (e.g., regionalization of the flow duration curve – FDC) (Castellarin et al. 2004). Further classification can be done within each group. For example, Castellarin et al. (2004) classified regionalization methods for FDC into statistical, parametric and graphical approaches. The methods used for estimating the time series of streamflows can be further categorized into three subgroups: a) model parameter estimation by developing regression relationships between model parameters and catchment characteristics (e.g., Magette ct al. 1976); b) transfer of model parameters, whereby a catchment similarity analysis is conducted and

---

[7] This chapter is based on the paper Regionalization of a conceptual rainfall-runoff model based on similarity of the flow duration curve: a case study from the semi-arid Karkheh Basin, Iran" by Masih, I.; Uhlenbrook, S.; Maskey, S.; Ahmad, M. D. 2010. *Journal of Hydrology* 391: 188-201. DOI:10.1016/j.jhydrol.2010.07.018.

parameters of gauged catchments are used in simulations for similar ungauged or poorly gauged catchment (e.g., Kokkonen et al. 2003; Wagener et al. 2007); and c) other regionalization techniques such as spatial interpolation of parameters (e.g., Merz and Blöschl 2004) or regional pooling of data for parameter estimation for ungauged catchments (e.g., Goswami et al. 2007).

Despite considerable progress in hydrological research, the prediction of streamflow for ungauged or poorly gauged catchments still remains a major challenge (Sivapalan et al. 2003; Wagener and Wheater 2006). A brief review of some key studies involving commonly used regionalization methods applying conceptual rainfall-runoff models for streamflow estimations in ungauged or poorly gauged catchments is presented in the following section. We defined a catchment as ungauged when no streamflow records exist, whereas a data limited or poorly gauged catchment is defined as a catchment where some measured streamflow records are available that are usually short, have many gaps and are of poor quality. These records are not enough to achieve a satisfactory level of model calibration for streamflow simulation.

5.1.2. Review of regionalization methods using conceptual rainfall-runoff models

An overview of some applications of the rainfall-runoff models for regionalization in different parts of the world is given in Table 14 and briefly discussed below. The selected studies estimated continuous time series of streamflows using a rainfall runoff model and reported the performance measures in terms of at least one of the three evaluation criteria, namely, Nash–Sutcliffe efficiency (NSE), coefficient of determination (R^2) and the mean annual volume balance (VB). These points were considered in the selection for consistency in comparison of this study and the presented literature in Table 14, Moreover, in selecting the literature for discussion we attempted to represent a wide range of hydro-climatic environments and provide reasonably good coverage of most of the regionalization methods.

Magette et al. (1976) used 21 catchments (0.02–12 km^2) in USA for regionalization of six selected parameters of the Kentucky Watershed Model (KWB). They used 15 catchment characteristics in developing regression equations and found that a multiple regression technique used in stepwise manner was successful in developing equations to estimate model parameters from catchment characteristics, but that simple linear regression models were totally unsuccessful. They randomly selected five out of 21 catchments for validation. Although the validation results showed significant variations, they concluded that the approach was useful and should be further developed. Vandewiele et al. (1991) used 24 catchments (16-2160 km^2) in Belgium for developing regression equations to estimate three parameters of a monthly conceptual rainfall-runoff model using the basin lithological characteristics. They concluded that their regionalization approach was capable of generating reliable monthly time series for ungauged sites within the region.

Servat and Dezetter (1993) evaluated the performance of two conceptual rainfall-runoff models (GR3 and CREC models) for possible applications to ungauged catchments in the north-western part of the Ivory Coast. They were able to relate all model parameters to catchment characteristics (rainfall and land cover) with varying degrees of success. The regionalization results in terms of R^2 and NSE were variable, particularly for the NSE which was quite low (i.e., close to zero) in some cases.

Ibrahim and Cordery (1995) applied a monthly water balance model for predicting streamflows in New South Wales, Australia. The used model had four parameters, of which three were estimated from rainfall data. Abdulla and Lettenmaier (1997) regionalized seven of the nine parameters of a large-scale model (VIC-2L) for Red and White river basins in USA. They estimated two of the model parameters from STATSGO soil data. For other parameters, they used 28 catchment variables, related to soil and climate, for developing multiple regression equations between model parameters and catchment variables. Their regionalization results were generally good in most cases, although they noticed better performance in humid and subhumid catchments than in semi-arid to arid catchments.

Seibert (1999) used the HBV model for a regionalization study using 11 catchments in Sweden and found that six of the 13 model parameters could be estimated from the land cover features (i.e., forest and lake areas). However, the application to ungauged catchments was achieved with varying degrees of success, with daily NSE ranging from 0.23 to 0.72. Merz and Blöschl (2004) compared eight regionalization methods using the HBV model with data sets from 308 catchments in Austria. Parajka et al. (2005) conducted a follow-up of the Merz and Blöschl 2004 study by improving the model structure (i.e., by dividing catchments into elevation bands of 200 m interval), adding snow cover data and conducting similarity analysis on the basis of catchment attributes. They concluded that the methods based on similarity approaches produce reasonably good regionalization results. This finding is also consistent with that of Kokkonen et al. (2003) who concluded that "*When there is reason to believe that, in the sense of hydrological behaviour, a gauged catchment resembles the ungauged catchment, then it may be worthwhile to adopt the entire set of calibrated parameters from the gauged catchment instead of deriving quantitative relationships between catchment descriptors and model parameters.*"

McIntyre et al. (2005) proposed a regionalization method of ensemble modeling and model averaging and tested it using a five parameter version of the probability distributed model (PDM) on 127 catchments (1-1,700 km^2) in the United Kingdom. They selected donor catchments based on catchment similarity analysis for which three catchment characteristics, i.e., catchment area, permeability and rainfall were used. In this approach more than one donor catchment is selected, which is different from the usual approaches of using a single donor catchment for streamflow simulations at an ungauged site. Then the full parameter set of each of the donor catchments is used to predict streamflows at the ungauged catchment, thereby, generating an ensemble of flow values. Then the average streamflow could be taken from the weighted average with weights defined based upon the relative similarity. They found that the proposed method performs reasonably well as compared to the established procedure of regressing parameter values from the catchment descriptors. However, they also noted that the new method estimated the low flows better than high flows. They recommended further testing of the model, especially to test different model types and improved definition of similarity.

Goswami et al. (2007) developed a methodology that uses a regionalization and multi-model approach for simulating streamflows in ungauged catchments. Like other methods, their methodology did not involve transfer of model parameters from gauged catchment to ungauged catchment, and model parameters need not be related to physical catchment descriptors. They used seven different models for regionalization and for each model three methods were tested that involved the use of the discharge series by taking regional averages, regional pooling of data and transposition of discharge data of the nearest neighbor. They used 12 gauged catchments in France to illustrate their methodology and each time considered one of them as ungauged for the application of the method and then compared the results with observed time series of daily discharge using the NSE criterion. The results indicated a mix of success and failure for the individual catchments and tested methods. However, they concluded that the pooling method of regionalization coupled with the conceptual soil moisture accounting and routing model (SMAR) was the best approach for simulating flows in ungauged catchments in that region. The second best method was the transposition of data from the nearest neighbor provided the catchments are similar in the hydro-meteorological, physiographic characteristics and drainage area.

Oudin et al. (2008) compared three widely used regionalization approaches for (a large number of) 913 French catchments (10-9,390 km^2) by using two conceptual rainfall-runoff models (GR4J and TOPMO models). They showed that regionalization based on the spatial proximity performed the best for their sample of catchments. They also noted that the dense network of tested catchments used in their study might have resulted in favor of spatial proximity approach and recommended that this approach should also be tested in other regions, particularly where less number of gauged catchments are available.

The presented studies reveal that considerable progress has been made to estimate streamflows at ungauged catchments and quite a number of promising methods have been developed over the past few decades. However, the studies also depict a mix of success and failure of the available methods within a study region or while comparing outcomes from the different regions. Moreover, the tested regionalization approaches indicate large variability in the achieved performance statistics, which shows considerable scope for further improvement. Therefore, there is every motivation to make further progress on this important subject of regionalization in hydrology.

Model Regionalization Based on the Flow Duration Curve

Table 14. An overview of some studies related to regionalization of conceptual rainfall-runoff models.

Country	Catchments	Drainage area, km²	Simulation time step	Evaluation measures for gauged and test catchments			Reference
				Volume balance, VB, mm/yr.	Coefficient of Determination, R^2 (-)	Nash-Sutcliffe Efficiency, NSE, (-)	
USA	16 (5)	0.04 to 12 (0.02 to 10)	Hourly	NA (-372 to 155)	NA	NA	Magette et al. 1976
Belgium	20(4)	16 to 2163 (73 to 148)	Monthly	-8 to 12 (-29 to 54)	NA	NA	Vandewiele et al. 1991
Ivory Cost	11 (5)	100 to 4500	Daily	NA	0.23 to 0.99 (0.62 to 0.99)	0.02 to 1 (0.02 to 0.45)	Servat and Dezetter 1993
Australia	18 (8)	10 to 1870 (156 to 1792)	Monthly	NA (-1 to 4)	0.73 to 0.94 (0.67 to 0.76)	0.69 to 0.94 (0.62 to 0.89)	Ibrahim and Cordery 1995
USA	34 (40)	168 to 5226 (442 to 6894)	Daily	NA (-11 to 134)	0.41 to 0.97 (0.05 to 0.81)	NA	Abdulla and Lettenmaier 1997
Sweden	11 (7)	7 to 950 (7 to 1284)	Daily	NA	NA	0.70 to 0.88 (0.23 to 0.72)	Seibert 1999
Austria	308 (308)	3 to 5000 (3 to 5000)	Daily	NA	NA	0.67 (0.32 to 0.56)*	Merz and Blöschl 2004
France	12 (11)	32 to 371 (32 to 371)	Daily	NA	NA	NA (-27.66 to 0.94)	Goswami et al. 2007

Notes:
Figures in parentheses correspond to the test catchments.
NA refers to information not available.
* Efficiency values refer to median of all 308 catchments during calibration phase and (in parenthesis) minimum and maximum median values of tested regionalization methods.

5.1.3. Scope and objective

The main research question examined in this paper is whether or not the parameters of a conceptual hydrological model applied to a gauged catchment can be successfully transferred for simulating streamflows in a hydrologically similar but data-limited or poorly gauged catchment. In this study, the HBV model (Bergström 1992) is used for streamflow simulations in the Karkheh River Basin, Iran. The hydrologic similarity is defined based on four measures, i.e., drainage area, spatial proximity, catchment characteristics and flow duration curve (FDC). FDCs are frequently used for comparing the response of gauged catchments, but their potential use for the regionalization of conceptual rainfall-runoff models for flow estimation for the poorly gauged catchments needs to be explored and is a main objective of this study. Streamflow data are required for the construction of an FDC. However, an FDC could be established from the catchment characteristics for ungauged catchments using available FDC regionalization methods (e.g., Castellarin et al. 2004). For poorly gauged catchments, the available records, though short, could be used for the FDC construction. These insufficient records may not be used directly for rainfall-runoff modeling as indicated in the previous section. Another limitation in their direct use for modeling purpose is the unavailability of other corresponding data sets required for modeling, e.g., climatic data for the same period as runoff data may not be available. These typical limitations were faced for the poorly gauged catchments in the Karkheh Basin providing the main motivation for this regionalization study.

The abovementioned methods evaluated in this study require very limited data resources and were most suitable in the context of the data-limited region under study. The other commonly used methods, such as regionalization of the model parameters, generally require data sets from a large number of gauged catchments for developing statistically sound relationships between model parameters and catchment characteristics. Due to limited availability of gauged catchments and necessary data sets, these data-intensive methods were not tested for the study area. Nevertheless, the results of this study were compared with those published in the literature from some widely recommended methods tested in other regions of the world.

5.2.   Materials and Methods

5.2.1   Study catchments and available data

In the Karkheh Basin streamflow data are not available for many catchments and the existing records have gaps. There were about 50 streamflow gauging stations installed after 1950 out of which only 24 have been measured continuously. Filling these data gaps by estimating missing streamflow time series for the poorly gauged catchments was required for a good understanding of the hydrology and its spatio-temporal variability, which in turn should guide informed water management decisions.

Eleven gauged catchments, draining tertiary-level streams (475-2,522 km^2), located in the upper mountainous parts of the Karkheh Basin were selected for this study (Figure 25 and Table 15).

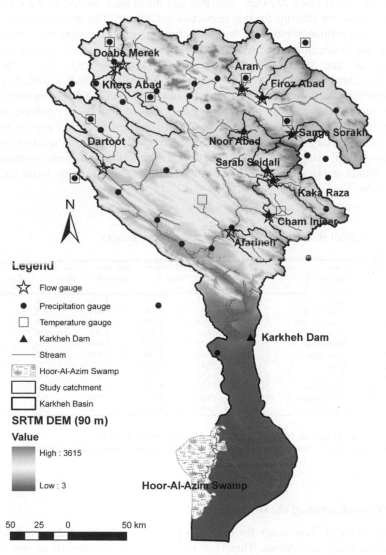

Figure 25. Salient features of the study area and location of the study catchments and used climatic stations.

The study period from January 1987 to September 2001 was selected considering the data availability/quality and representation of dry, wet and average climatic conditions. Time series of daily precipitation data for the study period were available for 41 climatic stations, well scattered across the study domain (Figure 25). The areal precipitation estimates were used in the model simulations, which were obtained by interpolation of the available station data by using inverse distance and elevation weighting (IDEW) technique (see chapter 6). Temperature data from eight climatic stations (Figure 25) were available and the station nearest to the catchment was used in the simulations for that respective catchment. The missing values in the data sets were estimated based on the values from neighboring stations. The missing values in the temperature data sets were few (less than 1% in most cases), with the exception of one station where records were available only for 1996-2001. Generally, temperature data of a station showed very good correlation with corresponding data from the neighboring stations ($R^2 > 0.90$) used for infilling of the missing records. In case of precipitation data, seven out of 41 stations had no missing records. On average, there were 5% in-filled precipitation events, ranging from 0 to 16%. Hargreaves equation (Hargreaves et al. 1985) was used to estimate the reference ET using daily data of maximum, minimum and mean temperatures. Further details on Hargreaves method and its application in the study basin are given in Appendix A.

Table 15. Salient features of the selected streamflow gauges.

Name of river	Name of station	ID	Long	Lat	Elevation, masl	Drainage area (km²)	Observed flow (mm/yr.)	Naturalized flow (mm/yr.)
Khorram Rod	Aran	1	47.92	34.42	1,440	2,320	59	87
Toyserkan	Firoz Abad	2	48.12	34.35	1,450	844	55	102
Gamasiab	Sange Sorakh	3	48.23	34.03	1,800	475	254	294
Qarasou	Doabe Merek	4	46.78	34.55	1,310	1,260	148	148
Abe Marg	Khers Abad	5	46.73	34.52	1,320	1,460	34	34
Bad Avar	Noor Abad	6	47.97	34.08	1,780	590	202	315
Abe Chinare	Dartoot	7	46.40	35.45	1,110	2,522	71	95
Chalhool	Afarineh	8	47.88	33.30	800	800	160	170
Khorramabad	Cham Injeer	9	48.23	33.45	1,140	1,590	223	341
Doab Aleshtar	Sarab Seidali	10	48.22	33.80	1,520	776	345	516
Har Rod	Kaka Raza	11	48.27	33.72	1,530	1,130	355	428

Notes: Long = Longitude (degrees East); Lat = Latitude (degrees North).
Data source: Ministry of Energy, Iran, with the exception of station ID and naturalized flow.

5.2.2   Naturalization of the streamflows

The abstraction of river water for irrigation purposes influenced the river flows in some of the study catchments. Therefore, naturalization of streamflows was carried out by adding abstraction rates, if any, to the observed streamflows. The main aim of doing naturalization of the streamflow was to improve the consistency of the regionalization procedures used in this study. The naturalization of the streamflows was considered helpful in reducing uncertainties arising due to the abstractions in the parameter estimation and consequent transfer from one catchment to the other. The direct pumping from the streams is the main mode of irrigation diversions by the farmers. However, no pumping records or data for other means of surface water diversions were available. Therefore, abstractions were estimated using the available information on crop ET, cropping patterns and cropped area, estimates of irrigation efficiencies and total annual abstractions. The procedure used is summarized below.

Calculation of crop water demand. The daily potential crop evapotranspiration (ET_c) was calculated using the following equation:

$$ET_c = \sum_{j=1}^{n} A_j Kc_j ET_0 \tag{10}$$

Where, ET_c is the total potential crop ET in m^3/d, A_j is the area under the jth crop in m^2, ET_o is the reference ET expressed in m/d estimated using Hargreaves method (Hargreaves et al. 1985), Kc_j is the crop coefficient for the jth crop (according to Allen et al. 1998), and n is the number of crop types, which are mainly wheat, barley, alfa alfa, sugar beat, maize and orchards. The data on cropping patterns and cropped area were obtained from JAMAB 1999 whereas sowing and harvesting dates were based on field surveys. The total ET_c was obtained by the summation of the values for the individual crops.

Calculation of irrigation demand and streamflow abstractions. The irrigation demand was estimated using the following equation:

$$I_d = ET_c \left(1 - \frac{e_p P}{ET_0} \right) \tag{11}$$

where, I_d is irrigation demand in m^3/d, P is the precipitation in mm/month and e_p (-) is the fraction of the precipitation effectively used as ET. The ratio of effective precipitation and reference ET was computed using monthly data of precipitation and ET_o. For the whole Karkheh Basin, JAMAB (1999) estimated that 66% of the annual precipitation is consumed as ET and 34% forms the renewable water resources. For this study conducted in the upper catchments of the Karkheh Basin, the value of e_p was assumed as 0.5, since the evaporation rates are lower in upper mountainous part of the basin compared to the lower arid plains.

The abstractions from the streams were estimated using the following equation:

$$I_{sw} = f_{sw} \frac{I_d}{\eta} \qquad (12)$$

where, I_{sw} is the surface water withdrawals, m^3/d, f_{sw} is the fraction of surface supplies in the total irrigation withdrawals and η (-) is the irrigation efficiency. The used values of η were in the range of 0.3 to 0.7 (JAMAB 1999). The lower values of η correspond to catchments with higher surface water withdrawals and vice versa. The annual values of f_{sw} were also available from the study of JAMAB (1999) who estimated total irrigation withdrawals from surface water and groundwater sources in the study catchments for the period 1993-94. The catchments where surface water was the main source of irrigation (i.e., $f_{sw} > 0.9$), the same value of f_{sw} was used for each day of the year. For catchments where conjunctive use of surface water and groundwater was present, the annual value of f_{sw} was distributed into monthly values following the supply-demand principle whereby higher values were assigned to the months having higher streamflows (i.e., March to June) and lower values to the months having lower streamflows (i.e., August to October). This way, f_{sw} was varied for each month but was kept constant for each day of a month. The estimated values of I_{sw} were compared with the available estimates at annual scale for the year 1993-94. If the difference was more than 15%, the procedure was repeated by modifying the values of η and monthly distribution of f_{sw}. The threshold of 15% was considered appropriate given the limitations related to the used data as well as full representations of the involved processes by (simplified) equations used in this method. Finally, I_{sw} values were added to the observed streamflow to get the naturalized streamflows. The observed and naturalized streamflows are given in Table 15, which indicates the extent of the influence of naturalization for each study catchment. As an example, Figure 26 shows the observed and naturalized streamflows of one catchment (Aran). This illustrates the streamflow differences in particular during the late spring and summer, when the crop water requirements are the largest. After discussions with local experts it was concluded that these corrections are reasonable and reflect the impact of local practices.

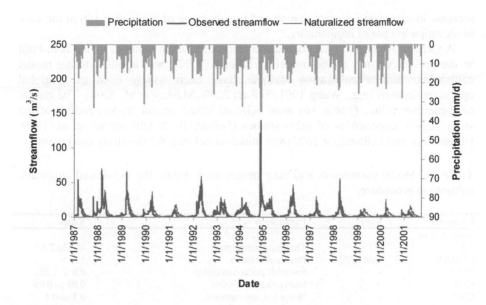

Figure 26. Naturalized and observed daily time series of streamflows of Aran catchment.

5.2.3. Model calibration and validation at the gauged catchments

The HBV model was applied to each of the 11 gauged catchments and was calibrated using daily climatic and streamflow data from January 1987 to September 2001. The data were split into calibration (October 1987 to September 1994) and validation (October 1994 to September 2001) periods. Before calibration, a warming-up period of 273 days was used for initialization so that model parameters attained appropriate initial values. Each catchment was divided into a number of elevation zones at an interval of 200 m. This interval was selected in order to balance the total number of elevation bands that could be accommodated in the HBV and SWAT (see next chapter) modeling set up. This threshold was also appropriate to avoid having too many or too less divisions of the study catchments. Each elevation zone was divided into three vegetation zones, namely forest (zone 1), cropland (zone 2) and range/bare lands (zone 3). Since the elevation is known to have major impacts on the distribution of rainfall and temperature, which have already been studied in the region, the values of the two parameters for lapse rates of precipitation and temperature were based on the earlier studies of Sutcliffe and Carpenter (1968), JAMAB (1999) and Muthuwatta et al. (2010). The values of lapse rates were kept constant for all catchments and set to an increase of 5.5% per 100 m increase in elevation for precipitation and to a decrease of 0.4 ^{0}C per 100 m increase in elevation in case of temperature.

A Genetic Algorithm (GA) based automatic calibration method, which is in-built in the present version of the model by Seibert (2002), was applied during model calibration. Similar calibration methods have been widely used as a global optimization tools (e.g., Wang 1991; Seibert 2000; Maskey et al. 2004). The ranges of parameter values (Table 16) were selected based on our understanding of the study region, experiences of other studies (Seibert 1999; Uhlenbrook et al. 1999; Uhlenbrook and Leibundgut 2002) and initial model runs for the study catchments.

Table 16. Model parameters and their ranges used during the GA-based automatic calibration procedure.

Parameter	Unit	Explanation	Range
Snow routine			
TT	^{0}C	Threshold temperature	-2.5 to 2.5
CFMAX	mm ^{0}C^{-1}d^{-1}	Degree-day factor	1 to 6
SFCF	-	Snowfall correction factor	0.8 to 1.25
CFR	-	Refreezing coefficient	0.05 to 0.05
CWH	-	Water-holding capacity	0.1 to 0.1
Soil routine			
FC	mm	Maximum of SM (storage in soil box)	50 to 500
LP	-	Evaporation reduction threshold (SM/FC)	0.5 to 0.7
BETA	-	Shape coefficient for soil storage/percolation	1 to 6
Response routine			
PERC	mm d^{-1}	Maximal flow from upper to lower box	0.1 to 6
UZL	mm	Threshold for Q_0 outflow in upper box	10 to 100
K_0	d^{-1}	Recession coefficient (upper in upper box)	0.05 to 0.5
K_1	d^{-1}	Recession coefficient (lower in upper box)	0.01 to 0.15
K_2	d^{-1}	Recession coefficient (lower box)	0.001 to 0.05
Routing routine			
MAXBAS	d	Routing, length of weighting function	1 to 5

For instance, the threshold temperature (TT) for snow was set to fall in the range of −2.5 to 2.5 ^{0}C. The optimized threshold value of this parameter defines whether the precipitation falls in the form of rain or snow. During winter months, the temperature may fall below the optimized snow temperature threshold causing precipitation to occur in the form of snowfall apart from the rain events during this period. The parameters of the snow and soil routines were estimated, using the abovementioned GA-based optimization procedure, in a distributed manner, thus having different values for each of the three vegetation zones. The parameters of the response and routing routines could only be estimated uniformly in the current version of the HBV model and were, therefore, representative of the whole catchment. The Nash-Sutcliffe Efficiency (*NSE*) estimated at the daily time step (equation 8) was used as an objective function to estimate the model performance (Nash and Sutcliffe 1970). The *NSE* is considered as a robust approach to assess the model goodness of fit in hydrological modeling and is widely used (e.g., ASCE 1993). However, it is also worth noting that the results based on *NSE* optimization could be biased towards high flows, which fact warrants caution in interpretations (e.g., Wagener et al. 2004). Other commonly used measures also have their own merits and constraints. For instance, the widely used performance measure, Coefficient of Determination (R^2), may reflect higher values (good performance) if the variability of two data sets is well synchronized despite their volumetric difference. Therefore, for having a better picture of the results, in addition to NSE, we examined R^2. The difference in the mean annual runoff, termed as volume balance (VB) was also examined.

5.2.4. Regionalization of model parameters based on catchment similarity analysis

In this study, the hydrological similarity was defined based on four similarity measures: drainage area, spatial proximity, catchment characteristics and flow duration curve (FDC). Once the similarity was established among 11 gauged catchments, the best parameter set of one catchment was transferred to another catchment (temporarily considered as ungauged, termed as *pseudo-ungauged*) for streamflow simulations. The whole parameter set was adopted from a donor catchment. The main advantage of adopting a complete parameter set is that the parameter interdependencies are not neglected. The results were then compared, in terms of NSE, R^2 and VB, by using the observed streamflow time series of the *pseudo-ungauged* catchment.

In terms of similarity in area, each of the 11 catchments was compared with other catchments and was rendered similar to the one which had the closest drainage area. Similarly for spatial proximity, the two catchments located nearest to each other were defined as similar. In cases where more than one catchment were available in the neighborhood, the catchment with the least distance from the centroid and/or having the longest common boundary was considered the most similar one. The similarity based on catchment characteristics was defined comparing the climate (ratio of mean annual precipitation and reference ET), topography (average catchment slope, elevation and stream density), land use (area under forest and crop land), soil (area under rock outcrop type soils) and geology (area under limestone-dominated geology). These characteristics are generally considered as the major drivers of the hydrological processes and catchment runoff response (Nathan and McMahon 1990b; Wagener et al. 2007). The similarity index (S) was calculated by using Equation (13) and the variables given in Table 17.

$$S = 1 - \sum_{i=1}^{M} \alpha_i \frac{\Delta V_i}{Max(\Delta V_i, \overline{V}_i)} \tag{13}$$

where, S is the similarity index (-) which takes a value between 0 and 1 and defines the degree to which catchment 1 is similar to catchment 2, M is the number of catchment characteristics (variables) used for computing the similarity index. The α_i are the weights (-) between 0 and 1 for the given characteristics such that sum of the weights is equal to 1. In this study, equal weights are used for all the characteristics. The variables V, ΔV, and $\overline{V}$ refer to the value of the respective catchment characteristics, the absolute difference between catchment 1 and 2, and the average value of catchment 1 and 2, respectively.

Table 17. Catchment characteristics used in calculating the similarity index.

Catchment		Catchment characteristics							
ID	Name	P/ET_o (-)	Slope (%)	Elevation (masl)	Stream density (km/km^2)	Rock outcrop soils (%)	Forest (%)	Cropland (%)	Limestone dominated geology (%)
1	Aran	0.292	15	1,768	0.061	54	10	48	52
2	Firoz Abad	0.292	17	1,949	0.063	56	10	30	27
3	Sange Sorakh	0.379	15	2,081	0.032	55	15	17	59
4	Doabe Merek	0.383	13	1,522	0.060	48	8	87	47
5	Khers Abad	0.312	10	1,529	0.076	49	10	73	20
6	Noor Abad	0.319	16	2,037	0.056	44	8	59	62
7	Dartoot	0.342	15	1,533	0.084	63	33	54	22
8	Afarineh	0.391	23	1,643	0.094	100	50	5	48
9	Cham Injeer	0.370	20	1,652	0.078	55	29	38	39
10	Sarab Seidali	0.353	27	2,100	0.061	71	8	45	61
11	Kaka Raza	0.357	23	2,024	0.084	63	13	34	60

In the fourth approach, similarity in the FDCs was compared both by means of visual inspection and by using a statistical criterion, Relative Root Mean Square Error (RRMSE). FDCs are very useful for comparing the hydrological response of catchments (e.g., Linsley et al. 1949; Hughes and Smakhtin, 1996; Yilmaz et al. 2008). Their shape is an indicator of catchment response to rainfall and also depicts the storage characteristics of the catchments and influence of topography, geology, vegetative cover and land use. In this study, the FDCs were plotted using daily discharge data which were normalized by the drainage area to facilitate comparison. The shape of the FDC for each catchment was visually compared with the FDCs of the other catchments; the catchments showing best match for both high and low flow percentiles were considered hydrologically similar. A commonly used objective criterion based on RRMSE, termed here as ε (–), Equation 14 was used to define the similarity between the FDCs.

$$\varepsilon = \frac{\sqrt{\dfrac{1}{N}\sum_{i=1}^{N}(Q_i - \hat{Q})^2}}{\overline{Q}} \tag{14}$$

where, Q_i is the ith flow percentile (mm/d) of one FDC and i ranges from 1 to N; $\hat{Q}_i$ is the corresponding ith flow percentile (mm/d) of another FDC; and $\overline{Q}$ is the mean discharge of the first (base line) FDC. The ε values were calculated for the whole FDC corresponding to the flow percentiles Q_0 to Q_{100} using daily discharge data.

5.2.5 Assessment of the impact of parameter uncertainty on the regionalization results

The issue of parameter uncertainty is well recognized in hydrological modeling (Uhlenbrook et al. 1999; Beven 2001; Wagener et al. 2004; McIntyre et al. 2005). Generally, parameter values are not unique, and results in large uncertainty bands in the discharge predictions. Furthermore, similar model simulations can be achieved by using different combinations of parameter values, which is generally termed in hydrology as equifinality or nonuniqueness of the model parameters (Beven 2001). In this study, the impact of parameter uncertainty on the regionalization results was also investigated. First, the best parameter set of a study catchment in the regionalization procedure was used, as indicated in the previous section. Then to check the consistency of the results, we selected 50 different parameter sets of a catchment that yielded the highest NSE values during the automatic calibration process, and used them for the regionalization in a way similar to that of using the single best parameter set. As mentioned in section 5.2.3, the automatic calibration was based on the GA-based optimization procedure. Therefore, the 50 best parameter sets are the ones resulting in the highest NSE out of the many good parameter sets that the GA-based optimization method generates. More parameter sets may be used for the purpose of this investigation, but we consider this number is reasonably good to test our hypothesis on the effect of parameter uncertainty on regionalization outcome. The regionalization results were considered reliable given the results remain consistent in terms of studied performance indicators (NSE, R^2 and VB) while using different parameter sets (e.g., both in case of the best parameter set and the 50 other good parameter sets).

5.3.   Results and Discussion

5.3.1. Model results of automatic parameter estimation

The calibration results showing the comparison of observed and simulated streamflows are provided in Table 18, summarizing the daily NSE, R^2 and VB estimates. The NSE values were quite good for most of the catchments (i.e., >0.6), with the exception of two catchments indicating values in the range of 0.41 to 0.46. Similar patterns were indicated by R^2 and VB, depicting reasonably good model performance in most cases. Although, during the validation period, NSE and R^2 values were lower than their corresponding values during the calibration period, the values were reasonably good in most cases (i.e., NSE >0.5). Furthermore, the performance results obtained in this study are in good agreement with those of other model regionalisation studies (e.g., Abdulla and Lettenmaier 1997; Merz and Blöschl 2004).

The calibration and validation results suggest that the optimized parameter sets could simulate the rainfall-runoff relationships reasonably well in most cases. However, it should be noted that the models are not perfect and may involve uncertainties resulting from uncertainties in the model structure, input data and parameter values (further discussed in section 5.3.3). Therefore, the results should be interpreted cautiously. For example, in the case of the Sange Sorakh (ID: 3) catchment the low performance was attributed to the high influence of groundwater discharge of a spring which the model was not able to simulate well given the high uncertainties in locating the boundaries of the karstified recharge area and complexity of the hydrological processes. The low performance of the Afarineh (ID: 8) could be mainly attributed possibly to high uncertainty in the climatic input data for this particular catchment due to less density of the climatic gauges in this area. In this catchment, the model consistently overestimated the average flows resulting in a high volume error and underestimated the high flood peaks.

Table 18. HBV model calibration and validation results, showing daily Nash-Sutcliffe efficiency (NSE), daily coefficient of determination (R^2) and annual volume balance (VB).

Catchment		Nash-Sutcliffe Efficiency (NSE, -)	Coefficient of determination (R^2 -)	Volume Balance (VB)		
ID	Name			Observed (mm/yr.).	Simulated (mm/yr.)	Difference (%)
Calibration						
1	Aran	0.91	0.91	95	90	-5
2	Firoz Abad	0.76	0.78	118	104	-12
3	Sange Sorakh	0.46	0.46	332	332	0
4	Doabe Merek	0.88	0.89	171	148	-13
5	Khers Abad	0.66	0.67	39	39	0
6	Noor Abad	0.64	0.70	349	326	-7
7	Dartoot	0.80	0.81	95	111	17
8	Afarineh	0.41	0.48	196	294	50
9	Cham Injeer	0.80	0.80	367	349	-5
10	Sarab Seidali	0.73	0.76	560	498	-11
11	Kaka Raza	0.83	0.84	483	405	-16
Validation						
1	Aran	0.67	0.81	79	95	20
2	Firoz Abad	0.45	0.64	85	94	11
3	Sange Sorakh	0.56	0.71	271	238	-12
4	Doabe Merek	0.66	0.69	129	96	-26
5	Khers Abad	0.68	0.69	30	37	23
6	Noor Abad	0.44	0.57	279	309	11
7	Dartoot	0.25	0.46	94	106	13
8	Afarineh	0.11	0.58	144	298	107
9	Cham Injeer	0.56	0.66	315	331	5
10	Sarab Seidali	0.59	0.68	471	450	-4
11	Kaka Raza	0.75	0.77	371	370	0

Notes:
Dartoot and Sange Sorakh had missing streamflow data. For Dartoot the calibration and validation results refer to the period October 1, 1994 to September 30, 2001 and October 1, 1990 to September 30, 1992, respectively. For Sange Sorakh the calibration and validation results refer to the period October 1, 1987 to September 30, 1994 and October 1, 1999 to September 30, 2001, respectively. For all other catchments the calibration and validation periods refer to October 1, 1987 to September 30, 1994 and October 1, 1994 to September 30, 2001, respectively.

5.3.2. Regionalization results based on drainage area, spatial proximity and catchment characteristics

The summary of the catchment similarity analysis is presented in Table 19, indicating most similar catchments whose parameters were transferred for the regionalization purpose under each of the four tested methods.

Table 19. Results of the catchment similarity analysis for the four tested methods.

Catchment		Catchment similarity based on the studied methods					
		Drainage area	Spatial proximity	Similarity index		Flow duration curve	
ID	Name	Similar catchment	Similar catchment	Similar catchment	Value of S	Similar catchment	Value of ε
1	Aran	Dartoot	Firoz Abad	Noor Abad	0.85	Firoze Abad	0.28
2	Firoz Abad	Afarineh	Aran	Aran	0.82	Aran	0.25
3	Sange Sorakh	Noor Abad	Sarab Seidali	Kaka Raza	0.70	Cham Injeer	0.37
4	Doabe Merek	Kaka Raza	Khers Abad	Noor Abad	0.81	Firoze Abad	0.84
5	Khers Abad	Cham Injeer	Dartoot	Doabe Merek	0.75	Aran	2.32
6	Noor Abad	Sange Sorakh	Sarab Seidali	Aran	0.85	Cham Injeer	0.29
7	Dartoot	Aran	Khers Abad	Cham Injeer	0.79	Aran	1.39
8	Afarineh	Sarab Seidali	Cham Injeer	Cham Injeer	0.67	Doabe Merek	0.99
9	Cham Injeer	Khers Abad	Kaka Raza	Dartoot	0.79	Noor Abad	0.27
10	Sarab Seidali	Afarineh	Noor Abad	Kaka Raza	0.83	Cham Injeer	0.39
11	Kaka Raza	Doabe Merek	Cham Injeer	Sarab Seidali	0.83	Sarab Seidali	0.61

The regionalization results for the calibration period are presented in Figure 27. The results of transferring the model parameters based on similarity in area show that in most cases the simulations were far away from the observed values in terms of NSE, R^2 and VB, with the exception of Kaka Raza (ID: 11) where the results were reasonably good. The regionalization based on spatial proximity showed much better simulations compared to those based on drainage area. Promising results were obtained for four catchments, namely, Aran (ID: 1), Firoz Abad (ID: 2), Doabe Merek (ID: 4) and Sarab Seidali (ID: 10), with NSE in the range of 0.51 to 0.78. But a large number of catchments resulted in poor simulations, i.e., four catchments had negative NSE values (ranging from -3.4 to -0.10). Similar to drainage area and spatial proximity, the regionalization results based on catchment characteristics were not better in most cases (Figure 27). Four out of 11 catchments produced comparatively better results with NSE and R^2 values in the range of 0.24 to 0.64 and 0.69 to 0.77, respectively. Rest of the catchments yielded poor results, particularly in terms of VB and NSE. On the whole, the results suggest that the above mentioned regionalization approaches are likely to produce unacceptable results in most cases. Therefore, none of them could be recommended for the regionalization purposes in the study region.

5.3.3. Regionalization results based on FDC

The FDC plots for all the study catchments are shown in Figure 28 and their similarities in terms of RRMSE (ε) are given in Table 19. In general, visual comparison and the used objective criteria indicated good correspondence with each other. Both visual comparison and ε values indicate that 7 out of 11 studied catchments revealed good similarity with at least one catchment in the study group. The ε values in these seven cases ranged from 0.25 to 0.61. The FDC-based regionalization results for these catchments were reasonably good, with five out seven catchments resulting in the NSE values in the range of 0.23 to 0.78 (Figure 27).

The R^2 values were also good, ranging from 0.54 to 0.87. Similarly, most of them depicted reasonably good performance in terms of VB. For instance, only two out of these seven catchments produced, negative NSE values, but still could simulate annual yields reasonably well (e.g., VBs for Sange Sorakh and Noor Abad were 1 and 24%, respectively). It is important to note that the Sange Sorakh catchment yielded lower NSE and R^2 values even during calibration. The lower performance, during calibration, validation and regionalization could be attributed to the significant contribution from a perennial spring, which the model was unable to simulate well given the high uncertainties in locating the geographical boundaries of the recharge area and the complexities in the hydrological processes in this region.

The FDCs of the remaining four catchments were not very similar to the rest of the study catchments. However, for consistency in the number of catchments used in all of the tested regionalization methods, we also executed FDC-based regionalization for these catchments by transferring the parameters from the catchment having the least value of ε. As expected, the results were not very good when compared to those catchments where similarity was adequately defined. Nevertheless, the outcome was comparable to the other three methods.

Furthermore, in most cases, the good regionalization results in case of tested methods other than the FDC-based method correspond to the pair of catchments having quite similar FDCs. For example, three out of four good performing catchments in case of spatial proximity (e.g., Aran, Firoz Abad and Sarab Seidali) also depicted similarity in the FDC of the corresponding neighbor.

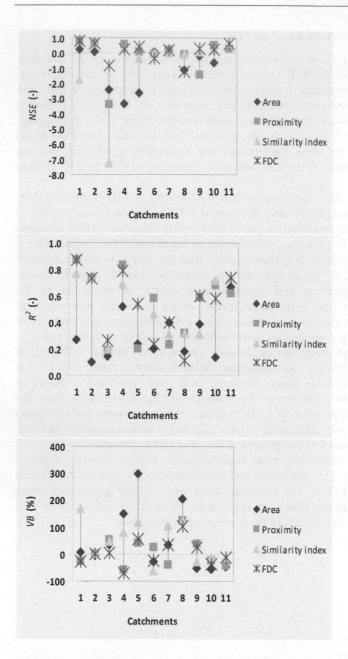

Figure 27.   Regionalization results of the four tested methods.
(The used catchment numbers in the x-axis correspond to the names as follow: 1: Aran; 2: Firoz Abad; 3: Sange Sorakh; 4: Doabe Merek; 5: Khers Abad; 6: Noor Abad; 7: Dartoot; 8: Afarineh; 9: Cham Injeer; 10: Sarab Seidali; 11: Kaka Raza.)

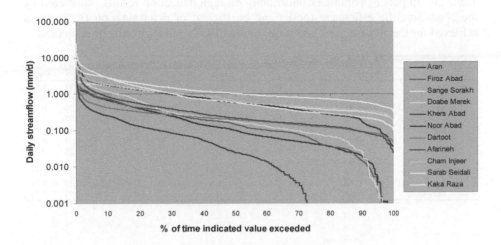

Figure 28. Comparison of FDCs for the similarity analysis.

5.3.4. Impact of parameter uncertainty on the regionalization results

The summary of the regionalization results using 50 best parameter sets for the FDC based regionalization method is presented in Table 20 and Figure 29. The resulting statistics given in Table 20 are reported in terms of median, 25th and 75th percentile, minimum and maximum. The presented statistics were obtained by arranging the results in descending order and then calculating various exceedance percentiles in a way similar to well-known flow duration analysis. This analysis helped to quickly view the degree of consistency when different parameter sets were used in the regionalization. For instance, if the range of different percentiles is small, then the impact of parameter uncertainty could be considered negligible. The results reveal that, despite different parameter sets, the regionalization results were reasonably consistent. This suggests that parameter uncertainty did not have considerable impact on the regionalization outcome. For example, maximum *NSE* values, achieved using the best parameter sets (as discussed in the previous sections 5.3.2 and 5.3.3) were not markedly different in most cases. This is further supported by the fact that the good-performing catchments continue to perform well for all of the 50 tested parameter sets (Table 20 and Figure 29). Moreover, none of the low performing catchments showed significant improvement as a result of using different parameter sets. The similar inferences were drawn regarding impact of parameter uncertainty on the regionalization results of the other three tested methods (not shown here).

Table 20. Impact of parameter uncertainty on regionalization results, illustrated by the Nash-Sutcliffe efficiency (NSE) and coefficient of determination (R2) results achieved for the 50 parameter sets used for the FDC-based regionalization method.

Catchment		Nash-Sutcliffe Efficiency (NSE,-)					Coefficient of determination (R^2, -)				
ID	Name	Median	P25	P75	Min	Max	Median	P25	P75	Min	Max
1	Aran	0.79	0.80	0.79	0.78	0.80	0.86	0.86	0.86	0.84	0.87
2	Firoz Abad	0.66	0.67	0.65	0.57	0.69	0.73	0.74	0.73	0.69	0.75
3	Sange Sorakh	-0.94	-0.80	-1.33	-1.77	-0.41	0.25	0.27	0.25	0.20	0.32
4	Doabe Merek	0.36	0.39	0.35	0.24	0.45	0.79	0.80	0.79	0.77	0.81
5	Khers Abad	0.26	0.44	0.11	-0.05	0.47	0.54	0.54	0.53	0.51	0.56
6	Noor Abad	-0.59	-0.48	-0.64	-0.88	-0.18	0.26	0.26	0.25	0.21	0.29
7	Dartoot	0.03	0.11	0.00	-0.11	0.26	0.27	0.28	0.26	0.25	0.59
8	Afarineh	-1.23	-1.22	-1.24	-1.31	-1.11	0.11	0.12	0.11	0.11	0.12
9	Cham Injeer	0.39	0.41	0.37	0.29	0.57	0.59	0.60	0.59	0.57	0.62
10	Sarab Seidali	0.26	0.28	0.23	0.16	0.31	0.61	0.62	0.59	0.55	0.63
11	Kaka Raza	0.59	0.60	0.58	0.56	0.62	0.73	0.74	0.73	0.71	0.74

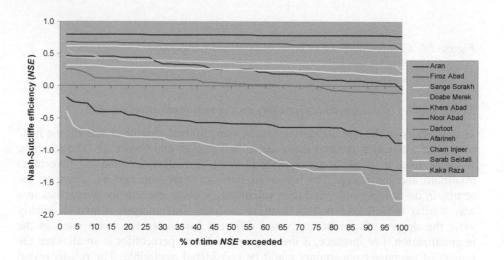

Figure 29. Impact of parameter uncertainty on regionalization results, illustrated by the exceeding percentiles of Nash-Sutcliffe efficiency (NSE) obtained from the 50 parameter sets used during regionalization based on similarity in the FDC.

5.3.5. Comparison of the FDC- based regionalization results with other studies

The results of this study indicate that the performance of the regionalization based on the similarity of the FDC is superior to that of the other three tested methods. Although, we could not test more methods due to limitations of the available data, we compared our findings with related studies conducted elsewhere using other methods. The comparison was made between the results of the FDC-based regionalization (Figure 27) with those of the studies presented in Table 14. The main aim of this comparison is to obtain an overview of the comparative position of the proposed FDC-based regionalization method among other widely recommended regionalization methods. Moreover, this comparison cannot replace a rigorous comparative assessment and is recommended as a future research activity. Therefore, it is acknowledged that this comparison should be interpreted cautiously because of inherent differences in the studies, i.e., differences in the amount and quality of the data sets used and varying hydro-climatic environments, among others.

The comparison reveals that the FDC-based regionalization approach stands very well among the most promising techniques developed elsewhere. For instance, the regionalization results based on the estimation of model parameters using catchment characteristics, indicated variable degrees of success, as demonstrated by the wide range of calculated performance measures (Table 14). The reported daily NSE values for the parameter regionalization studies of Servat and Dezetter (1993) and Seibert (1999) were in the range of 0.02-0.45 and 0.23-0.72, respectively. Similarly, the studies of Servat and Dezetter (1993) and Abdulla and Lettenmaier (1997) reported R^2 values in the range of 0.62-0.99 and 0.05-0.81, respectively. A similar trend of variable performance can be seen in many methods other than parameter regionalization. For example, Merz and Blöschl (2004) achieved median NSE values in the range of 0.32-0.56 for their eight regionalization methods tested for the 308 catchments, and Goswani et al. (2007) indicated NSE values in the range of -27.66-0.94 for their regional pooling method. The reported FDC-based regionalization results of this study for five out of seven catchments (where FDC similarity was well established) were in the range of 0.54-0.87 in terms of daily R^2 values and 0.23-0.78 in terms of daily NSE values. These encouraging results suggest that model regionalization based on the FDC similarity is a very good addition to the available regionalization methods.

However, all of the tested methods, including the FDC-based regionalization, resulted in some cases where the performance was not good. This suggests that the problem of achieving successful outcomes for all model applications for the poorly gauged or ungauged catchments still remains a challenging undertaking and, thus, needs further research. This could be attributed to the nature of the problem at hand as the degree of variability in the hydrological processes among different catchments is very high. Therefore, supporting the regionalization results through other sources of data and qualitative information is extremely desirable to avoid erroneous results. Nonetheless, the chances of invalid results drawn by applying the FDC-based regionalization method to poorly gauged catchments are likely to be small because at least some estimates of the streamflow characteristics are available for comparison in such cases (e.g., mean annual and monthly flows; various exceeding percentiles of flow).

5.4.   Concluding Remarks

This study examined the application of the HBV model for the generation of streamflow time series in data-limited catchments of the Karkheh Basin using model parameters transferred from similar gauged catchments. The similarities of the catchments for model parameter transfer were determined based on drainage area, spatial proximity, catchment characteristics and flow duration curve (FDC). Although the streamflow validation results based on spatial proximity and catchment characteristics are better than those based on geographical area, the overall results remain unsatisfactory in most cases. The study has shown that catchment similarity analysis based on FDCs provides a sound basis for transferring model parameters from gauged catchments to poorly gauged catchments in the Karkheh Basin. In most cases, the simulated time series of streamflows resulted in reasonably good values of the examined performance indicators (i.e., *NSE*, R^2 and *VB*) with negligible impact of the parameter uncertainty on the regionalization outcome. Furthermore, this new method also compares well with the studies conducted elsewhere using other promising methods. These demonstrations suggest that the new FDC-based regionalization method is a valuable addition to the available regionalization methods. The proposed method could be recommended for the practical applications for estimating time series of streamflows for the poorly gauged catchments in the mountainous parts of the Karkheh Basin. However, the poor performance in some cases for the promising regionalization methods indicates the complexity of the hydrological issues and of the regionalization problem and clearly highlights the scope for further improvements. This essentially requires more effort on better understanding the hydrology of ungauged or poorly gauged catchments and further developments in the regionalization procedures, in particular with regard to widely tested, and improving existing, methods, finding new regionalization approaches and exploring innovative ways of using available (scarce) data sets.

The methodology presented in this thesis is easy to replicate in other river basins of the world. Moreover, it can work well in the river basins, like the Karkheh Basin of Iran, facing a decline in streamflow monitoring networks and/or having a limited number of gauged catchments. Further testing of the proposed FDC-based regionalization method is highly recommended, i.e., by using different rainfall-runoff models, application under different hydro-climatic conditions, and for different extents of water resources development in the catchments (e.g., from more pristine to more regulated catchments).

6. IMPACT OF AREAL PRECIPITATION INPUT ON STREAMFLOW SIMULATIONS[8]

6.1. Introduction

The use of hydrological models in planning and management of water resources has become the norm, and a wide array of hydrological models (including freeware) is now available. The Soil Water Assessment Tool (SWAT) (Arnold et al. 1998; Neitsch et al. 2005; Gassman et al. 2007) is one such model. The main data sets required to formulate and run the model include a Digital Elevation Model (DEM), land use, soil, climatic and land use management data sets. The quality of these inputs has a significant impact on the model formulation process and on the results. Many studies have investigated the impact of the resolution of DEM, soil and land use data on the SWAT simulations (Chaplot 2005; Dixon and Earl 2009). Research has also been devoted to examine the impact of catchment subdivisions on the SWAT simulations (Jha et al. 2004; Tripathi et al. 2006). The studies on evaluating the impact of climatic data input on SWAT simulations (discussed below) are gaining increased attention, given the fact that climatic data are a major driver of hydrological and other processes simulated by the model. The current way of climatic data input in the SWAT is rather simplistic. Climatic data of a rain gauge located nearest to the centroid of a subcatchment are used for that subcatchment. This may not be accurate enough, particularly in regions where spatial heterogeneity is high (e.g., mountainous terrains), or where data are sparse but spatial variability of processes is not. This, in turn, has an impact on the model formulation process (e.g., parameterization) and quality of the simulated results (Oudin et al. 2006; Mul et al. 2009). For example, in response to over-predicted rainfall, the model parameterization process may tend to increase ET to match the observed and predicted streamflows. In many cases, finding the appropriate model structure and parameter sets may not work well in delivering acceptable model simulations if the input precipitation is inaccurate. Hence, improved precipitation input is very important to obtain good results (Oudin et al. 2006; Mul et al. 2009; Tobin and Bennet 2009).

---

[8] This chapter is based on the paper Assessing the impact of areal Precipitation input on streamflow simulations using the SWAT Model by Masih, I.; Maskey, S.; Uhlenbrook, S.; Smakhtin, V. 2011. *Journal of the American Water Resources Association* 47(1):179-195. DOI: 10.1111/j.1752-1688.2010.00502.x.

The impact of different spatio-temporal resolution of rainfall input on simulated runoff, using hydrological models other than SWAT, was examined in many studies (e.g., Faurès et al. 1995; Maskey et al. 2004; Tetzlaff and Uhlenbrook 2005). Although the results of these vary, they agree on the need to better represent precipitation input in modeling. A brief review of the studies attempted to address the issues of climatic data input in SWAT modeling is presented below.

Chaplot et al. (2005) studied the impact of rain gauge density on water, and sediment and nitrogen fluxes in two small catchments (51 and 918 km^2) in USA. Their study indicated that the use of higher rain gauge densities resulted in better simulations, particularly for sediment fluxes. Jayakrishnan et al. (2005) compared monthly and annual streamflows simulated by SWAT for the four catchments (196 to 2,227 km^2) in Texas, USA, by using rain gauge and radar (Next Generation Weather Radar, NEXRAD) data sets. They found that input of areal rainfall measured by radar performed better than that of the rain gauge data, despite some inherent limitations of the latter, particularly problems of accuracy at daily time scale. Watson et al. (2005) compared performances of three daily rainfall generation models using SWAT for a 308 km^2 catchment in Australia. They concluded that all three rainfall inputs produced reasonably good simulations of mean annual runoff. However, runoff variability was not well simulated given poor generation of rainfall variability. Cho and Olivera (2009) evaluated the impact of the resolution of land use, soil and precipitation data on simulated streamflows in three catchments (277 to 1005 km^2) in the USA. They formulated 18 models of each catchment by combining three land use, three soil types and two precipitation input scenarios. The two precipitation scenarios used were: a) using data from all available rain gauges, and b) using data from a single rain gauge for the whole catchment area. Each model was independently calibrated and validated. All models produced comparable values of daily Nash-Sutcliffe efficiencies. The main conclusion was that more refined representation of spatial data may not necessarily result in improved SWAT streamflow simulations in small catchments. Tobin and Bennett (2009) compared monthly streamflows simulated by SWAT using precipitation data collected through rain gauges, radar (NEXRAD stage III) and satellites (Tropical Rainfall Measurement Mission, TRMM) at the outlet of the two rivers in USA (Middle Nueces River catchment, 7,720 km^2, and the Middle Rio Grande River catchment, 8,905 km^2). Their findings revealed that streamflows were better simulated using radar data compared to the other two sources of precipitation input. Starks and Moriasi (2009) compared SWAT streamflow simulations using four resolutions of precipitation data on three experimental catchments (75 to 342 km^2). The number of rain gauges in three scenarios varied from one to seven. The fourth scenario used the radar precipitation data available at 4 km grid. The study indicated a satisfactory calibration of the SWAT model in all four cases, although the data set with higher rain gauge density and the radar-based precipitation produced comparatively better streamflow simulations.

These studies have strongly pointed out the need for more research on finding ways and means of improved precipitation input in SWAT simulations. Although previous investigations are very helpful steps in this direction, they remained limited in many features highlighting the need for further research. For instance, most of them were carried out in regions of USA which are generally considered data-rich compared to other countries, particularly the developing world (e.g., radar data are not available in many developing countries) and, therefore, remain limited to draw general conclusions. The studies represent small- to medium-sized catchments (50 to 9,000 km^2) and do not represent large river basins. Another important limitation is the lack of spatio-temporal coverage. The model performance has been evaluated at the catchment outlet in all cases, which prohibit explaining the spatial variations of the studied processes within a catchment. Similarly, few of the abovementioned studies using SWAT compared the performance at a daily time resolution, but were mostly limited to annual and monthly time scales. These shortfalls limit our understanding of the spatio-temporal impact of the improved input data on hydrological and other processes. Furthermore, these knowledge gaps hamper the informed basin-wide/regional planning and management of water resources (Santhi et al. 2008). Therefore, there is also a clear need for studies highlighting the spatio-temporal variability of the studied processes when comparing the impact of different sources of precipitation data on the model performance (Chaplot et al. 2005; Jayakrishnan et al. 2005; Watson et al. 2005; Cho and Olivera 2009; Starks and Moriasi 2009; Tobin and Bennett 2009).

The main research question addressed in this chapter is how improved precipitation input influences the hydrological simulations and, hence, impact water resources assessment across a large river basin. The specific objectives are: a) to compare the SWAT performance achieved by using different areal precipitation input, obtained by interpolation of the available rain gauge data and by using the rain gauge data as per SWAT's standard procedure; and b) to examine spatio-temporal performance of the model simulations under both precipitation input scenarios. The model was applied to the upper mountainous part of the Karkheh Basin (Figure 30a) covering an area of 42,620 km^2 from where almost all of the basin's runoff is generated (Figure 30b). The SWAT 2005 modeling system, version ARCSWAT 2.0 (Winchell et al. 2008) was used.

6.2. Data and Methods

6.2.1. Data used in the model setup

The Shuttle Radar Topography Mission (SRTM) Digital Elevation Model (DEM) of 90 m resolution was used for subcatchment definition. A drainage area of 300 km^2 was used as the threshold for the delineation of subcatchments. This threshold was chosen to balance the resolution of the available information. This way, the study area was divided into 71 subcatchments (Figure 30b). The delineated subcatchments were divided into different elevation bands using an elevation interval of 200 m. This helped account for the topographic impacts on the climate. The value of temperature lapse rate was set in the range of -2 to -5 ^{0}C/km, which is in close agreement with the estimated values for the study area (JAMAB 1999). The used values of precipitation lapse rate correspond to the annual lapse rate of 150-300 mm/km. This range is also in close agreement with earlier studies for the Karkheh Basin (Sutcliffe and Carpenter 1968; JAMAB 1999; Muthuwatta et al. 2010). The advantage of using ranges of precipitation and temperature lapse rates was to account for the likely differences in the topography and orographic impacts across the study basin. Since SWAT needs a daily precipitation lapse rate the annual values were translated into the required format using a procedure similar to that described by Fontaine et al. (2002). In the followed approach the annual values of lapse rates are distributed among all of the precipitation events in a year.

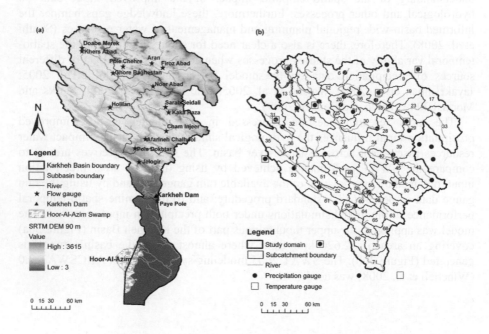

Figure 30. The Karkheh basin and location of the selected streamflow gauges (a); and the location of studied subcatchments and used climatic data stations (b).

The hydrological response units (HRUs) were defined based on information on land cover, soil and slope. The land cover map was prepared (Ahmad et al. 2009) using field data, GIS coverage and NDVI images based on remote sensing data from Moderate Resolution Imaging Spectroradiometer (MODIS) of 250 m resolution. It distinguishes 10 land use/land cover classes, with rain-fed farming (33%), forest (23%), rangelands (18%) and bare lands (15%) constituting about 90% of the study area. A digitized soil classification map was available from the Department of Soil and Water Research Institute (SWRI), Iran. This digital map had an original scale of 1: 1,000,000. The available soil map indicates fine to medium texture soils and rock outcrops (shallow soils) being the dominant soils in the study area.

The initial values of most of the model parameters for the soil classes were defined based on the results of the field tests conducted by the Iranian soil department at various field locations in the Karkheh Basin. Information was available about the soil texture, water content, soil depth, bulk density and some other parameters. The soil albedo information was defined based on Mathew's seasonal integrated surface albedo (Mathews 1983). The information on soil albedo was extracted from IWMI Integrated Data and Information System (IDIS) basin kit product for the Karkheh Basin (http://dw.iwmi.org/idis_DP/home.aspx). The other soil parameters were defined based on the SWAT soil data base, literature and field information. The topographic slope was derived from the DEM by using SWAT's HRU definition tool. The three categories of slope were defined to be used in the HRU definition, i.e., a) 0-8%; b) 8-30%; and c) > 30%. These slope categories represent level to undulating lands (0-8% slope), steep lands (8-30% slope) and mountains area (>30% slope) (FAO 1995). Finally, the HRUs were defined using the land use, soil and slope information. A threshold value of 5% for land use, soil and slope was used in the HRU definition. A threshold value of 5 to 10% is commonly used in HRU definitions to avoid small HRUs, reduce total number of HRUs and improve the computational efficiency of the model (Starks and Moriasi 2009; Tobin and Bennett 2009).

Daily climatic data for the period from January 1987 to September 2001 were used for the model simulations. Precipitation data from 41 stations and temperature data from 11 climatic stations were available. Locations of the used climatic gauges are shown in Figure 30b. The missing data were patched by using data of other stations based on a regression analysis. The study period was divided into a calibration period from October 1987 to September 1994 and a validation period from October 1994 to September 2001. In both periods, a warm-up period of 273 days was used to initialize the model.

6.2.2. Formulation of precipitation input scenarios

Climatic data of a station nearest to the centroid of a subcatchment was used in the model simulations as per SWAT's standard setup. This scenario of station precipitation input is hereafter referred to as Case I. The interpolated precipitation data were used as the model input in the second scenario (Case II). The inverse distance and elevation weighting (IDEW) technique was used for the interpolation of the available station data (see next section). The resulting precipitation was aggregated at the subcatchment level. Then a virtual precipitation gauge having the interpolated catchment precipitation was assigned for each of the 71 subcatchments. In this scenario (Case II), model simulations were performed by changing the precipitation data but keeping the rest of the data and the model structure the same as in Case I. The formulated SWAT models for both scenarios were independently calibrated using an automatic calibration procedure, discussed in detail under the following section on the model calibration.

The model performance was evaluated at 15 streamflow gauges. The selected gauges had a catchment area in the range of 590-42,620 km^2 (Figure 30a and Table 21). These stations were well distributed across the Karkheh River system. The studied gauges represent the primary- (Saymareh and Karkheh rivers), secondary- (Gamasiab, Qarasou, and Kashkan rivers) and tertiary-level streams. Finally, a comparison was made between the model performance achieved by Case I and Case II. The hydrological performance was evaluated using the Coefficient of Determination (R^2) and the Nash-Sutcliffe Efficiency (NSE) (Nash and Sutcliffe 1970) measures at daily and monthly time scales (equations 13 and 14, respectively). The relative difference in the observed and simulated mean annual streamflows was also compared.

Table 21. Geographical characteristics of the selected streamflow gauges in the Karkheh Basin.

Name of river	Name of station	Long	Lat	Elevation (masl)	Drainage area (km^2)	Number of rain gauges (No.)	Rain gauge density, (1 station per km^2)
Khorram Rod	Aran	47.92	34.42	1,440	2,320	2	1,160
Toyserkan	Firoz Abad	48.12	34.35	1,450	844	1	844
Gamasiab	Pole Chehre	47.43	34.33	1,280	10,860	11	987
Qarasou	Doabe Merek	46.78	34.55	1,310	1,260	2	630
Abe Marg	Khers Abad	46.73	34.52	1,320	1,460	2	730
Qarasou	Ghore Baghestan	47.25	34.23	1,268	5,370	7	767
Har Rod	Kaka Raza	48.27	33.72	1,530	1,130	2	565
Doab Aleshtar	Sarab Seidali	48.22	33.80	1,520	776	1	776
Khorramabad	Cham Injeer	48.23	33.45	1,140	1,590	1	1,590
Chalhool	Afarineh	47.88	33.30	800	800	1	800
Kashkan	Pole Dokhtar	47.72	33.17	650	9,140	6	1,523
Bad Avar	Noor Abad	47.97	34.08	1,780	590	1	590
Saymareh	Holilan	47.25	33.73	1,000	20,863	20	1,043
Karkheh	Jelogir	47.80	32.97	450	39,940	32	1,248
Karkheh	Paye Pole	48.15	32.42	125	42,620	32	1,332

Notes: Long = Longitude (degrees East) and Lat = Latitude (degrees North).
Data source: Ministry of Energy, Iran, barring the last two columns.

Preparation of the precipitation input for Case II

The earlier studies for the Karkheh Basin have demonstrated that topography has a strong influence on the spatial distribution of precipitation in this mountainous region (Sutcliffe and Carpenter 1968; JAMAB 1999; Muthuwatta et al. 2010). Elevation is known to be an important factor governing the spatial variability. These findings are in general agreement with those of other mountainous regions of the world (e.g., Daly et al. 2002). Moreover, the rain gauge data may not adequately represent the precipitation over an entire catchment. This issue is further exacerbated for catchments where rain gauge density is lower, such as for the region under study. Under such conditions, areal precipitation is likely to represent catchment conditions better compared to the station data.

For Case II, the daily station data were interpolated and aggregated at the subcatchment level using the IDEW technique. The hydrological data processing software called HyKit developed at UNESCO-IHE was used (Maskey 2007). The distance weighting method has already proven to perform well as compared to some other standard methods of perception regionalization for the Karkheh and its neighboring basins in the Zagros mountains, Iran (Saghafian and Davtalab 2007). The method has also been successfully used in other regions of Iran (Modallaldoust et al. 2008). The HyKit also offers the possibility of defining elevation weighting along with the distance weighting, making it more suitable for mountainous regions where topographic impacts on precipitation are important. The mathematical form of the equation used for interpolation is as follows:

$$\hat{p}_k = W_D \sum_{i=1}^{N} \frac{1}{D} w(d)_i \, p_i + W_Z \sum_{i=1}^{N} \frac{1}{Z} w(z)_i \, p_i \qquad (15)$$

where, $\hat{p}$ in mm per time step is the interpolated precipitation for a grid cell, W_D (-) and W_Z (-) are the total weighting factors for distance and elevations, respectively, p_i is the precipitation value in mm per time step of the ith gauge station and N is the number of gauges used in the interpolation for the current grid cell. Similarly, $w(d)_i$ (-) and $w(z)_i$ (-) are the individual gauge weighting factors for distance and elevation, respectively, and D (-) and Z (-) are the normalization quantities given by the sum of individual weighting factors $w(d)_i$ and $w(z)_i$, respectively, for all the gauges used in the interpolation. The weighting factors $w(d)_i$ and $w(z)_i$ based on inverse of distance and elevation, are given by the following equations:

$$w(d) = 1/d^a \quad \text{for } d > 0 \qquad (16)$$

$$w(z) = \begin{cases} 1/z_{\min}^b & \text{for } z \le z_{\min} \\ 1/z^b & \text{for } z_{\min} < z < z_{\max} \\ 0 & \text{for } z \ge z_{\max} \end{cases} \qquad (17)$$

where, d is the distance in km between the current grid and the gauge station used for interpolation, z is the absolute elevation difference (expressed in m) between the current grid cell and the gauge station used for interpolation, a and b are exponent factors for distance and elevation weightings, respectively. The exponents (a and b) and the abovementioned weighting factors are dimensionless numbers and z_{min} and z_{max} (expressed in m) are the minimum and maximum limiting values of elevation differences for computing elevation weightings (Daly et al. 2002). The use of z_{min} helps avoid the dominance of the stations having very small elevation difference (e.g., 10s of meters) from the target cells. The typical value of z_{min} varies from 100 to 300 m. The limit on maximum elevation difference enables data point inclusion to be restricted to a local elevation range. A typical z_{max} ranges from 500 to 2,500 m.

Note that in this interpolation, no grid cell contains more than one gauge station and that the grid cell which contains a gauge station will retain the same precipitation as that of the gauge station. The main advantage of distance weighting technique is its simplicity and ease of application to large data sets (e.g., daily time series). The inclusion of elevation weighting is helpful for improving the results in the mountainous regions where elevation could play a major role in the precipitation distribution. The method also has some limitations that mainly relate to the careful choice of the sensitive parameters.

Daily time series of precipitation from all of the 41 available gauges were used for interpolation in 5×5 km^2 grids, which are then aggregated to subcatchments as defined in the SWAT model. The parameters used in the interpolation were defined, primarily based on recommendations from the available literature and carrying out a cross validation exercise. The final parameter values were: $a = 2$, $b = 1$, $d = 70$, $W_D = 0.8$, $W_Z = 0.2$, $z_{min} = 100$ m and $z_{max} = 1,500$ m. The used parameter values were in good agreement with the literature (Daly et al. 2002). The interpolated results were cross-validated at 10 selected rain gauge locations/grid cells. The validation was done using the Jack-knife cross-validation approach (Quenouille 1956). In this method, interpolation runs were carried out for each of the 10 validation stations using data of all other stations excluding the current validation station (e.g., 41-1 = 40 in this case). Then, the interpolated and observed data for that station were compared by estimating R^2 between them. The interpolated values were in good agreement with the observed ones. The mean and standard deviations of monthly R^2 were 0.91 and 0.04, respectively. As expected, the daily R^2 values were comparatively lower than the monthly ones (with mean R^2 of 0.62 and standard deviation of R^2 of 0.13). However, considering high spatial variability of precipitation in this mountainous terrain, the achieved R^2 values were considered satisfactory.

Moreover, a correlation analysis among the rain gauge stations was also carried out to further evaluate the used radius of influence. The analysis indicated that most of the stations falling within a distance of 70 km exhibited a good correlation with one another (e.g., greater than 0.8 at monthly time scale). The 70 km radius of influence ensured the use of 2-15 stations in the interpolation for a subcatchment. Generally, the interpolation used more stations in the subcatchments located in the upper part of the study area because of the high station density compared to the middle and lower parts (Figure 30b).

6.2.3. Model calibration

The main options used in the SWAT model set up included: a) Soil Conservation Services Curve Number (SCS-CN) method, with crack flow not active, for estimating surface runoff (Soil Conservation Service Engineering Division 1986), b) Hargreaves method for daily potential evapotranspiration calculation (Hargreaves et al. 1985), and c) variable storage method for water routing in the streams (Williams 1969).

The formulated SWAT models for Case I and Case II scenarios were independently calibrated using an auto-calibration procedure. The SWAT-CUP software was used for this purpose (Abbaspour 2008). The Sequential Uncertainty Fitting algorithm (SUFI-2) was applied for the parameter optimization (Abbaspour et al. 2004; 2007). The SUFI-2 optimization follows 9 major steps, discussed in detail by Abbaspour et al. (2007), which are enumerated below.

1. An objective function is selected from the given options (e.g., R^2 or NSE etc.).
2. Physically meaningful ranges of the parameters are defined. Generally, wide ranges are suggested at this first step, which are revised in the following rounds of analysis.
3. A sensitivity analysis is performed to get a first hand view of the sensitive parameters.
4. Initial uncertainty ranges are assigned to parameters for the first round of Latin Hypercube sampling.
5. A Latin Hypercube sampling is carried; leading to n parameter combinations, where n is the number of desired simulations.
6. The simulations are assessed by estimating the objective function values.
7. A series of measures (e.g., sensitivity matrix) is calculated to evaluate each sampling round.
8. Measures assessing the uncertainties are calculated.
9. Because parameter uncertainties are initially large, the value of uncertainty measures tends to be quite large during the first sampling round. Hence, further sampling rounds are needed with updated parameter ranges. In this step, new parameter ranges are suggested, which are generally narrower than those defined in step 2. Then the whole procedure is repeated until desired results on parameter optimization are achieved.

This computationally efficient procedure is being increasingly used in the recent SWAT applications (e.g., Faramarzi et al. 2009) and is known to produce comparable results with widely used auto-calibration methods (Yang et al. 2008).

The parameters were optimized using R^2 as the objective function. Using a procedure similar to that adopted by Faramarzi et al. (2009), the SWAT model was simultaneously calibrated using daily streamflow data of the abovementioned 15 streamflow gauging stations. Hence, the best parameter set was the one which produced maximum value of the average R^2. The NSE was not used as an objective function in this study, mainly due to the possibility of a badly simulated station (with a large negative value) dominating the optimization process. However, the results were also evaluated in terms of NSE and annual volume balance (VB), in addition to R^2. Using more than one performance evaluation measure was considered helpful in evaluating the robustness of the calibration process. This was also useful in compensating for the specific limitation of a specific performance evaluation criterion.

The SWAT-CUP offers the possibility of selecting an objective function from the six different available options (e.g., sum of squares, R^2, weighted R^2 associated with slope-termed as bR^2, and NSE). Each of the abovementioned performance measure has its own merits and limitations (e.g., ASCE 1993; Krause et al. 2005; Gupta et al. 2009). Generally, these objective functions tend to better fit the simulated hydrographs to the high flows to achieve a higher value of the objective function, which often comes at the expense of relatively poor simulation of the low flows (e.g., Krause et al. 2005). The use of most widely applied and well- recommended performance measures, i.e., R^2, NSE and VB, was considered appropriate for the purpose of this study.

Before applying auto-calibration, a rigorous manual calibration exercise was performed. This helped in defining suitable initial values/ranges of the parameters, which were based on information from various sources that included measured data, global data sources, the SWAT soil and land cover database, literature, discussion with the local experts and field visits. For instance, the initial values of the parameters of the snow routine were defined in a way to obtain resultant snowfall values in good agreement with a recent study in the region (Saghafian and Davtalab 2007). Moreover, the used parameter values/ranges were in line with the literature (Fontaine et al. 2002; Jones et al. 2008). The SCS curve number values were varied for each of the land use categories and the used values were in close agreement with the literature (Soil Conservation Service Engineering Division 1986). Similarly, the selected parameters of the groundwater routine were spatially varied based on the information available from earlier studies (JAMAB 1999; Tizro et al. 2007; Masih et al. 2009). The finally used values and/or ranges resulted from the auto-calibration procedure, for both scenarios are presented in Table 22.

Table 22. Appropriate values and/or ranges of the selected parameters used in setting up SWAT model for the cases I and II.

Parameter	Suggested ranges in SWAT	Mode of change during auto-calibration runs[a]	Final value or ranges used for Case I	Final value or ranges used for Case II	Parameter sensitivity indicated by t value[b]
Snowfall temperature, SFTMP (°C)	-5-5	v	2.8	1.06	0.71
Snowmelt temperature, SMTMP (°C)	-5-5	v	1.2	0.16	0.66
Maximum melt rate of snow during a year (occurring in summer solstice), (mm °C^{-1} d^{-1})	0-10	v	1.9	4.3	0.46
Minimum snowmelt rate during a year (occurring in winter solstice), (mm °C^{-1} d^{-1})	0-10	v	1.5	2.1	0.81
Snowpack temperature lag factor (TIMP)	0-1	v	0.7	0.96	1.22
Minimum snow water content that corresponds to 100% snow cover, SNOCOVMX (mm)	0-500	v	127	492	2.12
Snow water equivalent that corresponds to 50% snow cover	0-1	v	0.3	0.2	6.10
Soil available water capacity (SOL_AWC), mm/mm	0-1	r	0.06-0.21	0.11-0.41	1.06
Soil saturated hydraulic conductivity (SOL_K), (mm/hr)	0-2000	r	4.5-36	4.8-38.4	7.79
Maximum Soil Depth (SOL_ZMX), (mm)	0-3500	r	150-2400	118-1180	3.76
Soil evaporation compensation factor, ESCO	0-1	v	0.61	0.87	0.003
Plant uptake compensation factor, EPCO	0-1	v	0.78	0.65	0.56
Soil albedo, SOL_ALB	0-0.25	r	0.09-0.11	0.08-0.10	0.04
Soil bulk density, SOL_BD (g/cm³)	0.9-2.5	r	1.70-1.99	1.62-1.82	3.10
Curve number, CN	30-100	r	57-90	56-88	3.20
Manning's n value for overland flow	0.01-30	r	0.02-1.61	0.01-0.94	1.53
Surface runoff lag coefficient (SURLAG)	1-24	v	1	4.65	6.41
Maximum canopy storage, CANMX (mm)	0-100	r	0-4.65	0-4.85	1.29
Base flow recession constant (ALPHA_BF)	0-1	r	0.1-0.3	0.08-0.24	2.98
Groundwater delay from soil to groundwater, (GW_DELAY), (d)	0-500	r	9-129	24-144	5.96
Fraction of total aquifer recharge percolated to deep groundwater, RCHRG_DP	0-1	r	0.07-0.70	0.08-0.75	1.57
Manning's n value for the main channel, CH_N2	-0.01-0.3	v	0.19	0.21	17.64
Effective hydraulic conductivity in the main channel alluvium, CH_K2 (mm/hr)	-0.01-500	v	2.26	1.11	0.53

Notes:

[a]v refers to the absolute change in the parameter made by replacing a parameter by a given value; r refers to the relative change in the parameter made by multiplying the parameter by 1 plus the factor in the given range (Abbaspour et al. 2007)

[b]t-value indicates parameter sensitivity; higher values refer to more sensitive parameters and vice versa.

The overall automatic calibration process done for this study consisted of three SWAT-CUP iterations, each composed of 250 simulations. As per the SWAT structure some of the parameters can only be defined for the whole study area and therefore will only have one global value for all subcatchments (e.g., parameters of the basin file including snow routine and SURLAG). The sensitivity analysis of the selected parameters was also conducted using the sensitivity analysis tool included in the SWAT-CUP. This helped understand the relative importance of the selected parameters for the study region. The results are presented in Table 22, indicating the relative sensitivity of each parameter. The results of the sensitivity analysis are generally in agreement with the literature (van Griensven et al. 2006; Faramarzi et al. 2009; Tobin and Bennet 2009).

It is important to recognize here that good model simulations can be achieved using various combinations of the model parameters. Therefore, the calibrated parameter values given in Table 22 do not necessarily represent the uniquely best parameter combination. This issue is well comprehended in hydrology and often termed as equifinality or nonuniqueness of the parameters (Uhlenbrook et al. 1999; Beven 2001).

6.3.   Results and Discussions

6.3.1.   Comparison of precipitation input

Figure 31a presents the mean annual precipitation under Case II. The results showed substantial variations in annual totals ranging from 370 to 640 mm/yr. Generally, the subcatchments located in the northeast, central and southern parts of the study area depict comparatively lower precipitation whereas, the catchments located in the southeast parts of the study basin have the highest precipitation. The topography seems to be the major driver of these spatial variations. Westerly winds are the main source of moisture in the study area (Demroes et al. 1998), which are strongly influenced by the topographic features causing high spatial variability. In general, the presented pattern of spatial variability is in good agreement with the available precipitation maps of the region (Saghafian and Davtalab 2007; Muthuwatta et al. 2010).

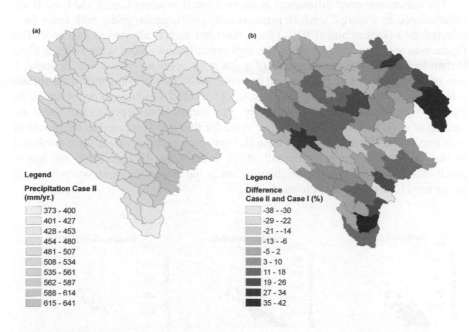

Figure 31. Mean annual precipitation for Case II (a) and percentage difference between Case II and Case I (b).

Figure 31b shows the percentage difference in the mean annual precipitation in Case II compared with Case I. The comparison revealed significant differences, indicating both increases and decreases in the range of -38 to 42%. The overall precipitation dynamics in Case II could be different from Case I in a number of ways. This is illustrated by the comparison of daily, monthly and annual precipitation for three selected subcatchments (Figure 32). Since SWAT simulates streamflow and other processes at daily time scale, it is more important to closely examine the differences in daily precipitation to understand the impact on simulated results. The daily values in Case II could be higher, lower or similar as compared to Case I. However, they show a clear pattern in extreme values. Generally, lower precipitation events can be totally missed out by a single rain gauge. These events are better accounted for in Case II. This is shown in Figure 32 by the daily precipitation values extending up to 20 mm/d in Case II (along the x-axis) when Case I shows no precipitation (zero value for the y-axis as indicated by some of the data points falling on the x-axis line). The high precipitation extremes in Case II are comparatively smaller in most cases when compared to Case I, though they could be the other way round for some subcatchments and events because these outcomes largely depend upon the influence of the neighboring gauge records.

The abovementioned differences in the precipitation under Case I and Case II are substantiated by Figure 33, which presents daily precipitation under both cases for a selected area (subcatchment ID: 1) for a short but significant period in a year. The figure indicates that, in most cases, the high precipitation values were lower in Case II than in Case I, with the exception of a few with opposite results. For example, 1 mm of precipitation was recorded on March 2, 1996 by the rain gauge used for the selected area. In contrast, neighboring gauges recorded a quite high precipitation on that day (up to 52 mm). Consequently, the interpolated precipitation based on records of several stations had a higher value for that day compared to the record of only one station. Similarly, under Case II, some precipitation values were attributed to the days which show no rainfall at all under Case I. This could be due to precipitation occurring in the neighboring areas but not on the area where the rain gauge used under Case I was located.

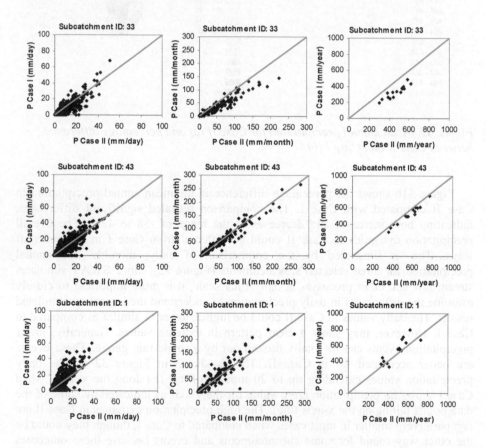

Figure 32. Comparison of daily, monthly and annual precipitation among Case I and Case II, illustrated by three selected subcatchments.

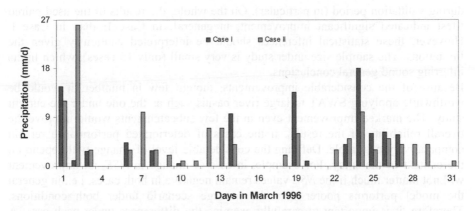

Figure 33. Comparison of daily precipitation for Cases I and II for a selected month (March 1996).

6.3.2. Comparison of streamflow simulations

The summary of the results on the studied performance measures are presented in Table 23a and 23b. Scatter plots of *NSE* and R^2 are presented in Figure 34, providing a quick overview of the comparison between Case I and Case II. Moreover, the magnitude of changes in daily *NSE* and R^2 in Case II compared to Case I is reflected in Figure 35. The monthly differences follow a pattern similar to that of the daily results (not shown in Figure 35 but could be inferred from Table 23a and 23b). The results indicate variable performance under both cases, indicating both increases and decreases. Nonetheless, better results were obtained under Case II in most cases, in particular for the smaller catchments. In general, streamflow regimes of the Karkheh River and its major tributaries were modeled reasonably well under both cases during calibration as well as validation periods. For example, monthly *NSE* values during calibration and validation periods for the two selected gauges on the Karkheh River, Jelogir and Paye Pole, were 0.76-0.89 and 0.77-0.91 under Case I and Case II, respectively. The corresponding *NSE* values at daily time scale were 0.69-0.83 and 0.65-0.81.

A paired *t*-test was applied to investigate the statistical significance of the observed differences on the whole. The significance of the resultant test statistics is noted in Table 23a and 23b. The results depict significant improvement for the *NSE* under Case II compared with Case I both at daily and monthly time resolutions, indicated by significant results at 95% confidence level. Similarly, *NSE* values under Case II were significantly better under both calibration and validation periods than under Case I.

Significant improvements in monthly R^2 during the calibration period were noted, but in the rest of the cases R^2 values were comparable under both scenarios. The simulated volume balance was significantly better in Case II than in Case I

during validation period (in particular). On the whole, the results of the used paired
t-test indicated significant improvement, in general, in Case II than in Case I.
However, these statistical inferences should be interpreted cautiously given the
limitations. The sample size under study is very small (only 15 cases), which limits
inferring sound general conclusions.

Because of the considerable improvements, though few in number, it would be
worthwhile applying SWAT to large river basins such as the one under the current
study. The marked improvement even in the few subcatchments would improve the
overall reliability of the results, if the cases of deteriorated performance remain
comparatively negligible. Defining the considerable level of change will depend on
the context of the study. For example, in case of change in *NSE*, an improvement
will not matter much if the *NSE* values remain negative in both cases, i.e., in general
the model performs poorer than the average scenario under both conditions.
Therefore, it is important to carefully examine the differences under each case. A
further discussion on the cases with similar, improved and deteriorated performance
is presented below.

Table 23a.   Comparison of the model simulations during calibration under Case I and Case II.

ID	Name of Gauge	Drainage area (km^2)	Calibration (October 1987 to September 1994)						
			Daily		Monthly		Mean annual flow (m^3/s)		
			R^2	NSE	R^2	NSE	Observed (m^3/s)	Simulated (m^3/s)	Difference (%)
Case I									
1.1	Aran	2,320	0.74	0.72	0.87	0.85	5.0	4.1	-18
1.2	Firoz Abad	844	0.51	0.17	0.72	0.55	1.9	2.1	11
1.3	Pole Chehre	10,860	0.82	0.75	0.91	0.88	41.5	37.6	-9
2.1	Doabe Merek	1,260	0.61	0.36	0.81	0.71	6.7	7.1	6
2.2	Khers Abad	1,460	0.45	-0.12	0.71	-0.21	1.8	2.6	43
2.3	Ghore Baghestan	5,370	0.81	0.81	0.90	0.89	25.3	23.3	-8
3.1	Kaka Raza	1,130	0.65	0.47	0.73	0.50	14.7	8.3	-44
3.2	Sarab Seidali	776	0.68	-0.95	0.87	0.25	9.6	10.6	10
3.3	Cham Injeer	1,590	0.78	0.64	0.90	0.76	12.6	12.8	1
3.4	Afarineh	800	0.51	-0.01	0.63	-1.36	4.7	7.9	67
3.5	Pole Dokhtar	9,140	0.81	0.78	0.92	0.87	64.7	64.5	0
4.1	Noor Abad	590	0.35	-0.58	0.57	-0.09	4.5	3.6	-21
4.2	Holilan	20,863	0.83	0.81	0.91	0.90	86.7	74.6	-14
4.3	Jelogir	39,940	0.87	0.83	0.95	0.89	184.5	181.8	-1
4.4	Paye Pole	42,620	0.76	0.74	0.93	0.89	209.6	183.2	-13
Case II									
1.1	Aran	2,320	0.77	0.77	0.90	0.90	5.0	5.2	4
1.2	Firoz Abad	844	0.52	0.48	0.71	0.71	1.9	1.9	-1
1.3	Pole Chehre	10,860	0.81	0.80	0.90	0.88	41.5	36.6	-12
2.1	Doabe Merek	1,260	0.65	0.63	0.90	0.88	6.7	5.4	-19
2.2	Khers Abad	1,460	0.51	0.14	0.81	0.02	1.8	2.7	48
2.3	Ghore Baghestan	5,370	0.82	0.80	0.93	0.91	25.3	21.0	-17
3.1	Kaka Raza	1,130	0.79	0.63	0.88	0.67	14.7	9.9	-33
3.2	Sarab Seidali	776	0.66	-0.24	0.84	0.54	9.6	8.7	-10
3.3	Cham Injeer	1,590	0.78	0.58	0.90	0.75	12.6	13.1	4
3.4	Afarineh	800	0.52	0.52	0.62	0.56	4.7	4.6	-3
3.5	Pole Dokhtar	9,140	0.79	0.78	0.93	0.91	64.7	56.4	-13
4.1	Noor Abad	590	0.53	0.33	0.72	0.60	4.5	4.3	-5
4.2	Holilan	20,863	0.84	0.82	0.92	0.90	86.7	73.7	-15
4.3	Jelogir	39,940	0.83	0.81	0.93	0.91	184.5	177.1	-4
4.4	Paye Pole	42,620	0.72	0.70	0.91	0.88	209.6	179.1	-15
Paired t-test significance[a]			NS	**	**	**	NA	**	NS

Note:

[a] The significance of the paired *t*-test refers to: NS: not significant; **: significant at 95% confidence level; NA: not applicable

Table 23b.   Comparison of the model simulations during validation under Case I and Case II.

ID	Name of Gauge	Drainage area (km²)	Validation (October 1994 to September 2001)						
			Daily		Monthly		Mean annual flow (m³/s)		
			R²	NSE	R²	NSE	Observed (m³/s)	Simulated (m³/s)	Difference (%)
Case I									
1.1	Aran	2,320	0.73	0.30	0.83	0.40	3.7	5.3	42
1.2	Firoz Abad	844	0.40	-2.48	0.49	-1.72	1.0	1.9	85
1.3	Pole Chehre	10,860	0.76	0.71	0.86	0.80	28.3	31.3	10
2.1	Doabe Merek	1,260	0.23	0.08	0.30	0.21	4.7	3.9	-18
2.2	Khers Abad	1,460	0.67	-0.12	0.88	-0.40	1.3	2.3	73
2.3	Ghore Baghestan	5,370	0.73	0.72	0.80	0.79	17.0	15.2	-11
3.1	Kaka Raza	1,130	0.69	0.61	0.84	0.74	10.7	7.8	-27
3.2	Sarab Seidali	776	0.67	-1.35	0.73	-1.06	7.3	9.2	26
3.3	Cham Injeer	1,590	0.61	-0.63	0.72	-0.75	9.9	13.2	34
3.4	Afarineh	800	0.41	-0.94	0.73	-1.68	3.4	7.0	106
3.5	Pole Dokhtar	9,140	0.73	0.56	0.88	0.52	47.1	60.8	29
4.1	Noor Abad	590	0.57	0.37	0.71	0.34	3.0	3.7	21
4.2	Holilan	20,863	0.86	0.84	0.88	0.87	63.7	58.9	-8
4.3	Jelogir	39,940	0.87	0.74	0.93	0.76	140.1	156.6	12
4.4	Paye Pole	42,620	0.73	0.69	0.85	0.78	167.7	160.1	-5
Case II									
1.1	Aran	2,320	0.72	0.63	0.81	0.70	3.7	5.3	43
1.2	Firoz Abad	844	0.50	-0.04	0.57	0.11	1.0	1.6	57
1.3	Pole Chehre	10,860	0.72	0.71	0.82	0.82	28.3	30.1	6
2.1	Doabe Merek	1,260	0.49	0.48	0.70	0.66	4.7	3.5	-26
2.2	Khers Abad	1,460	0.56	-0.29	0.88	-0.07	1.3	2.3	75
2.3	Ghore Baghestan	5,370	0.71	0.70	0.84	0.83	17.0	14.9	-12
3.1	Kaka Raza	1,130	0.74	0.67	0.85	0.78	10.7	9.0	-16
3.2	Sarab Seidali	776	0.68	-0.06	0.73	0.15	7.3	7.4	1
3.3	Cham Injeer	1,590	0.67	-0.20	0.80	-0.15	9.9	12.4	26
3.4	Afarineh	800	0.55	0.44	0.70	0.62	3.4	4.5	33
3.5	Pole Dokhtar	9,140	0.74	0.72	0.89	0.82	47.1	51.8	10
4.1	Noor Abad	590	0.54	0.35	0.74	0.40	3.0	3.8	23
4.2	Holilan	20,863	0.85	0.84	0.88	0.88	63.7	58.6	-8
4.3	Jelogir	39,940	0.82	0.78	0.90	0.84	140.1	148.2	6
4.4	Paye Pole	42,620	0.66	0.65	0.79	0.77	167.7	151.0	-10
Paired t-test significance[a]			NS	**	NS	**	NA	**	**

Note:

[a] The significance of the paired t-test refers to: NS: not significant; **: significant at 95% confidence level; NA: not applicable

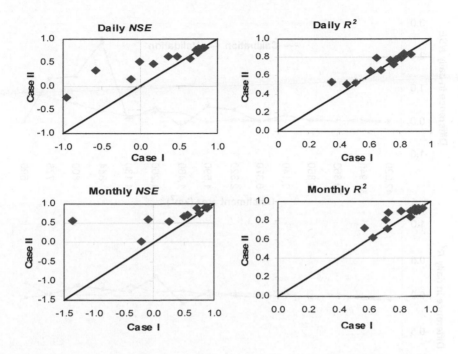

Figure 34. Scatter plots of NSE and R^2, highlighting the comparative performance under cases I and II.

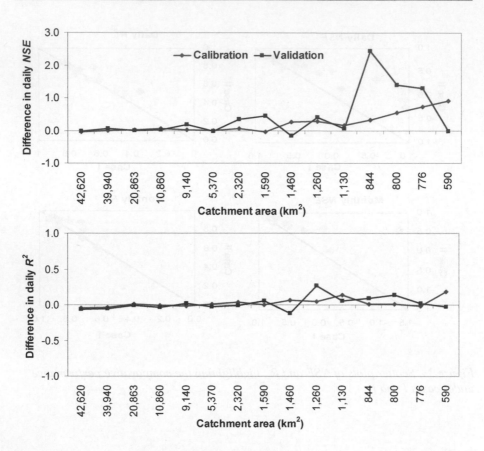

Figure 35. difference in the daily NSE and R^2 in Case II as compared to Case I for the calibration and validation periods.

Stations indicating good performance in both cases

The simulated results corresponded well with the observed values under both cases in six out of 15 studied flow gauges. These stations are: Paye Pole, Jelogir, Holilan, Pole Chehre, Pole Dokhtar and Ghore Baghestan. The daily *NSE* during the calibration period in Case I and Case II ranged from 0.74 to 0.83 and from 0.70 to 0.82, respectively. The corresponding values during the validation period were in the range of 0.56-0.84 and 0.65-0.84, respectively. Furthermore, there were no marked differences in the values of R^2 and *NSE* in all of these gauges, though with the exception of Pole Dokhtar where a noteworthy improvement of 0.3 in the monthly *NSE* for the validation period was observed. All of these stations represent primary- and secondary-level streams. This suggests that the change in precipitation input had minimal impact for larger catchments (in general). Daily hydrographs (observed and simulated) for the Jelogir station at the Karkheh River are shown in Figure 36a, as an example, indicating an almost similar pattern of simulations under both cases.

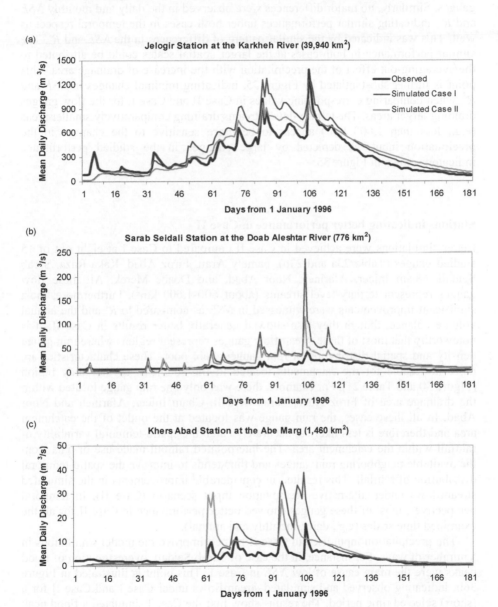

Figure 36. Observed and simulated daily hydrographs for Cases I and II for a selected period January to June 1996 at three stations: (a) Jelogir, (b) Sarab seidali, and (c) Khers Abad.

Furthermore, there were no striking differences in the performance changes in Case II when compared with Case I with respect to the spatial location of these gauges. Similarly, no major differences were observed in the daily and monthly NSE and R^2, indicating similar performances under both cases in the temporal respect as well. This was indicated by the similar pattern of differences in the NSE and R^2. The similar performance in both cases at the larger spatial scales could be attributed to the averaging-out effect of the precipitation with the increase of drainage area. This point is further substantiated by Figure 35, indicating minimal changes in NSE and R^2, when comparing corresponding values in Case II and Case I, for the flow gauges draining larger areas. The tertiary-level streams draining comparatively smaller areas (e.g., less than 1,600 km^2) appear to be more sensitive to the changes in the precipitation input, as depicted by large changes in the studied performance indicators shown in Figure 35.

Stations indicating better performance in Case II

Better simulations were achieved in Case II compared to Case I at eight out of 15 studied gauges (Table 23a and 23b), namely Aran, Firoz Abad, Kaka Raza, Sarab Seidali, Cham Injeer, Afarineh, Noor Abad, and Doabe Merek. All these flow gauges represent tertiary-level streams (about 600-1,600 km^2). Furthermore, most significant improvements were witnessed in NSE as compared to R^2 and the annual volume balance, though they also showed generally better results in Case II. It is noteworthy that most of these streamflow gauges represent regions where rain gauge density and spatial distribution of rain gauges were poor. These characteristics are more noticeable for the catchments located in the southeast of the study region (Figure 30 and Table 21). For example, there was only one rain gauge located within the drainage area of Firoz Abad, Sarab Seidali, Cham Injeer, Afarineh and Noor Abad. In all these cases, the rain gauge was located at the outlet of the catchment area and therefore is less likely to accurately catch the spatio-temporal variability of rainfall within the catchment area. The interpolated rainfall made use of data from the available neighboring rain gauges and thus tends to improve the spatio-temporal distribution of rainfall. This resulted in considerable improvements in the simulated streamflows under alternative precipitation input scenario (Case II). In temporal perspectives, most of these gauges showed better performance in Case II for all the examined time scales (e.g., daily, monthly and annual).

The precipitation input used in Case II helped improve the model simulations in a number of ways. For instance, in the case of Sarab Seidali, overestimation of flood peaks were the main cause of low NSE in Case I. This point is illustrated in Figure 36b, indicating observed and simulated streamflows under Case I and Case II for a (short) selected time period. The results show that the Case I simulated a flood peak of 237 m^3/s on April 14, 1996 (Julian day: 105) against the observed flow of 32 m^3/s for that day. Examination of precipitation events showed that higher precipitation was observed for the gauge used for this catchment compared to neighboring gauges. The used gauge in Case I recorded a precipitation value of 63 mm/d for

April 14. But the neighboring gauges recorded comparatively lower precipitation values, less than 40 mm/d for that day. This suggests that this precipitation event triggered this very high peak flow. However, the large difference in observed and simulated flows during this event could be attributed to very high rain in a small localized area as recorded at the gauge used for Sarab Seidali. The records from neighboring stations indicate that the used station precipitation did not represent the whole catchment well. The areal average of precipitation used in Case II considerably helped better simulate this peak flow, though it still did not perfectly match the observed flow on that day. However, the generally better estimation of flood peaks during the calibration and validation periods under Case II as compared to Case I could be attributed as the main reason for the improved performance under Case II for this catchment.

The other major reason for the better simulations in Case II could be the better representation of the overall precipitation amounts besides good depiction of the temporal dynamics of the individual precipitation events. For instance, in the case of Kaka Raza the station nearest the centroid recorded consistently lower precipitation volumes compared to the neighboring stations. The precipitation totals and the overall rainfall dynamics were better represented after interpolation (Case II) that, in turn, enhanced the accuracy of the simulated results. On the other hand, stations used in Case I for the catchment gauged at Afarineh observed higher precipitation compared to its neighbors causing poor simulated results. For this catchment, the annual volume balance during calibration and validation periods under Case I was 67 and 106% when compared with the observed streamflows. The simulated values in Case II considerably reduced this departure from the observed records and resulted in a volume balance of -3% (calibration) and 33% (validation). Furthermore, NSE values were also better in Case II than in Case I. The daily NSE under Case I were -0.01 and -0.94 during calibration and validation periods, respectively. On the other hand, the corresponding NSE results in Case II were 0.52 and 0.44, respectively.

Stations indicating good performance in Case I

No station performed considerably better in Case I than in Case II. However, there were two catchments, Firoz Abad and Khers Abad, where simulated results were generally poor in both cases, as indicated by inferior values of R^2, NSE and volume balance (Table 23a and 23b). In the case of Khers Abad, the percentage difference in the mean annual runoff was considerably higher in both cases, i.e., 43% under Case I and 45% under Case II during calibration and 73% under Case I and 75% under Case II during validation. Under both cases, R^2 and NSE ranged from 0.45 to 0.67 and -0.40 to 0.14, respectively. These points are substantiated by Figure 36c, indicating comparison of the observed and simulated discharge under Case I and Case II for the Khers Abad catchment. The poor performance could be due to a number of reasons. The input precipitation may not be well represented in both cases. However, considering comparatively higher number of stations located within this catchment and close to it, precipitation may not be a major reason for lower performance in this case. The observed mean annual water yield at Khers Abad (about 30 mm/yr.) is very low compared to the neighboring areas. This was unusual when considering largely similar climatic and physiographic patterns in this and neighboring catchments. During field observations and discussion with the local experts, it was noted that surface water-groundwater interactions were very complex and poorly known in this region. The area is highly influenced by karst formations and there are many springs in the neighboring regions. Therefore, the incoming precipitation might be heavily recharged to the deep groundwater/karst formations, which could become the source of spring flows outside of the topographic boundary of this catchment. These aspects were considered very important for proper modeling of this region and warrant in-depth scientific investigations. In the case of Firoz Abad, streamflows were heavily influenced by the water withdrawals for irrigation uses, and this was considered as the major factor for poor performance, besides uncertainties in other input data including precipitation under both of the tested scenarios. The poor performance of the hydrological models applied to the heavily regulated catchments is generally in line with the literature (e.g., Faramarzi et al. 2009).

6.4. Concluding Remarks

This study compared the SWAT model performance under a) standard SWAT precipitation input procedure (using records of the station nearest to the centroid of a subcatchment - (Case I) and b) modified areal precipitation input obtained through spatial interpolation (Case II). The model performance was assessed by using R^2, NSE at daily and monthly temporal resolutions and by comparing mean annual runoff at 15 selected streamflow gauges located in the Karkheh Basin.

The results show that, in general, the model performance was almost similar in both cases when evaluated in terms of R^2. However, a notable improvement was observed in the NSE criterion in Case II compared to Case I for eight out of 15 studied gauges (600 to 1,600 km^2). For these catchments, the performance in terms of R^2 and annual volume balance in Case II was either comparable or better when compared with Case I. Most of these catchments represent regions with limited climatic data, i.e., either rain gauge density was comparatively low or the distribution of the rain gauges within the catchment was poor (e.g., used gauge in Case I was often located at the outlet of the catchment). The improvement in the simulated streamflows in Case II was attributed to the improved representation of the precipitation regime and its spatial variability.

However, the results from both cases were comparable, in terms of all the studied performance measures, for the gauges located on the larger Karkheh River and its major tributaries with drainage areas larger than 5,000 km^2. Furthermore, for these gauges, no significant differences could be identified in terms of their spatial location (e.g., among the streams draining the upper, middle or lower parts of the study area) or between the studied time scale (e.g., daily, monthly and annual). The similar performance at the large spatial extent under both precipitation input procedures could be attributed to the averaging-out effect of the precipitation input while simulating the hydrological processes.

It can be concluded that the use of areal precipitation, obtained through interpolation of the available station data, improved the SWAT model simulated streamflows in the study basin. The results were strongly influenced by the spatial extent of the investigations as well as by the station density and spatial distribution of the available rain gauge data used in the interpolation. Further testing of (semi-) distributed hydrological models such as SWAT, using areal precipitation as an input (e.g., obtained through interpolation of rain gauge records, radar data and satellite observations), for its added value to the streamflow simulations and other processes is highly recommended. Future investigations should focus on the spatio-temporal aspects of the hydrological simulations, particularly in the large river basins, and should also study the impact of calibration strategies, i.e., by using various objective functions for the parameter optimization and performance evaluation, and by following different calibration, validation and uncertainty analysis procedures. Development of an optional component for the interpolation of climatic data within the existing SWAT model will benefit multiple SWAT users.

Although this study focuses on improvement of precipitation input in the SWAT model, the procedures and results are instructive for rainfall-runoff modeling in general.

7. QUANTIFYING SCALE-DEPENDENT IMPACTS OF UPGRADING RAIN-FED AGRICULTURE[9]

7.1 Introduction

Improvements of rain-fed agriculture are required to ensure global food security. Improved rain-fed agriculture also contributes to the global poverty reduction as the majority of the world's rural poor depend on rain-fed agriculture for livelihoods. It is also beneficial for environment, e.g., to reduce soil erosion. Yet, a proper understanding of trade-offs resulting from such interventions is essential too (e.g., CAWMA 2007; Rockström et al. 2010).

Wakindiki and Ben-Hur (2002) conducted a field-scale evaluation of the indigenous soil and water conservation techniques in a semi-arid rain-fed region of Kenya and concluded that the techniques they investigated, i.e., building trash lines of various sizes and materials, significantly reduced soil erosion and improved crop yields. The study also noted significant reduction in the surface runoff under the studied techniques. Makurira et al. (2010) suggested that the food and livelihood security of the farmers in semi-arid to arid regions could be significantly improved by promoting rainwater harvesting. Their field scale experimentations conducted in the Makanya Basin, Tanzania, demonstrated that the combined use of conservation agriculture, diverting runoff onto field plots and enhancement of in-field soil moisture through trenching and soil bunding (locally called *fanya juus*) could help in managing erratic distribution and scarce quantity of rainfall. The study showed that these methods could significantly increase plant transpiration resulting in higher crop yields and water productivity. Oweis and Hachum (2009) reported examples of successful implementation of various water harvesting techniques (e.g., contour ridges, semi-circular and trapezoidal bunds, small runoff basins, terraces, wadi-bed cultivation and tanks) from West Asia and North Africa region. They reported that the widespread adoption of water harvesting and supplementary irrigation techniques helped improve land cover growth and raise productivity levels, but required careful evaluation of factors such as available technical skills at the local level, characterization of climate, water and land use systems, prevailing institutional and policy environment and possible conflicts in the water uses and users among upstream-downstream areas.

---

[9] This chapter is based on the paper Quantifying scale-dependent impacts of upgrading rain-fed agriculture in a semi-arid basin by Masih, I.; Maskey, S.; Uhlenbrook, S.; Smakhtin, V. 2011. *Agricultural Water Management* (in review)

Xiubin et al. (2003) compared the observed runoff and precipitation records for two periods, representing hydrological conditions without implementation of soil and water management interventions (1959-1969) and the period (1990-1995) with the interventions in the three subbasins of the Yellow River Basin, China. They noted a reduction of about 50% in the mean annual runoff, which was mainly attributed to various interventions, such as building earth dams, planting trees or grass, terraces, and irrigation projects. They highlighted that the benefits of increased food production and reduced soil erosion realized from the abovementioned interventions came at the cost of reduction in the downstream flows. Lacombe et al. (2008) investigated the impact of water and soil conservation works (WSCW), mainly contour ridges and hillside reservoirs, on runoff response of the Merguellil Basin (1,183 km^2) in Tunisia. The observed rainfall and runoff records over 1981 to 2005 were used to investigate the changes in the runoff regime. The study indicated runoff reduction of 28-32% due to WSCW at the basin scale. They further noted that harvested soil moisture and stored water in the small dams were not efficiently used for the benefit of increased crop production, and argued that the adopted WSCW contributed to the loss of water through enhanced (non-beneficial) *ET* in the region.

The brief review of the recent studies presented above shows the need for much better understanding of the impact of upgrading rain-fed agriculture on hydrology and water availability at subbasin to basin scale. The main objective of this chapter is to investigate such impacts in the semi-arid Karkheh Basin, Iran. More specifically, this study aims to a) investigate the potential for upgrading rain-fed agriculture to irrigated agriculture and associated impacts on streamflow, b) evaluate the impact of soil and water conservation on streamflow, and c) assess the predictive uncertainty of the model used and its implications.

7.2.   Methodology

7.2.1.   Model used for the scenario simulation

A semi-distributed process-based hydrological model Soil Water Assessment Tool (SWAT) (Arnold et al. 1998; Gassman et al. 2007) was used to simulate various scenarios (discussed below). The model covers an area of 42,620 km^2 up to the outlet of the study basin at Paye Pole (Figure 30). We adopted the model with areal precipitation input obtained through interpolation, as it performs better compared to the standard SWAT precipitation input procedure of using data of a rain gauge located nearest the centroid of a subcatchment (see chapter 6). The modeling details such as on the input data, parameterization, calibration and validation can be found in chapter 6. An assessment of prediction uncertainty of the model was carried out in this study, and is discussed below.

The uncertainty analysis was carried out using the Sequential Uncertainty Fitting algorithm (SUFI-2) (Abbaspour et al. 2007) available in the SWAT-CUP software

(Abbaspour 2008). In this approach, all uncertainties, i.e., that of model, parameters and input data, are mapped on to the model parameter ranges. In SUFI-2, the prediction uncertainty of the model is evaluated using two measures: *P*-factor and *R*-factor. The *P*-factor which may vary from 0 to 100% indicate the percentage of observed data falling within the 95 percent prediction uncertainty (95PPU) band calculated at 2.5 and 97.5 percentiles of the model output based on Latin hypercube sampling. The *R*-factor is the average width of the 95PPU band divided by the standard deviation of the observed data.

The results presented in Figure 37 show the simulated streamflows along with the uncertainty band and the observed data. The results are shown here for six selected stations averaged over the 13 years period from 1988 to 2000. These stations were selected to keep consistency with the study objectives, with a focus on basin to subbasin scale impacts. The Paye Pole station at the Karkheh River (outlet of the study domain corresponding to river reach ID 71) reflects basin level implications of tested scenarios. The changes in streamflows at this location are pivotal to understand the water availability for the multipurpose Karhkeh Dam and its downstream area. The other selected stations are important to reflect the spatial variations within the basin, and represent the Karkheh River and its all major subbasins. Locations of these gauges are marked in Figure 6, and some basic features are given in Table 3.

The calibration and uncertainty results shown in Figure 37 reveal that most of the observed streamflow data fall well within the model's prediction uncertainty band, with about 2 months falling slightly outside of the 95PPU band in most cases. The *P*- and *R*- factors on daily values are also reasonably good (e.g., *P*-factor > 0.5, and *R*-factor <0.5). Similarly, the Nash-Sutcliffe Efficiency (*NSE*) (Nash and Sutcliffe 1970) and co-efficient of determination (R^2) on daily streamflows are also in good range (*NSE*: 0.65-0.80; R^2: 0.66-0.81).The percentage difference between observed and simulated mean annual flows was also quite small in most cases ranging from -13 to 8%. The reported performance statistics (*NSE*, R^2 and volume balance) were estimated from the simulations of the baseline simulation against the observed flows, which represented the final parameter set adopted in this study. The achieved performance statistics compare very well with other SWAT applications in Iran (Faramarzi et al. 2009) and elsewhere (e.g., Gassman et al. 2007).

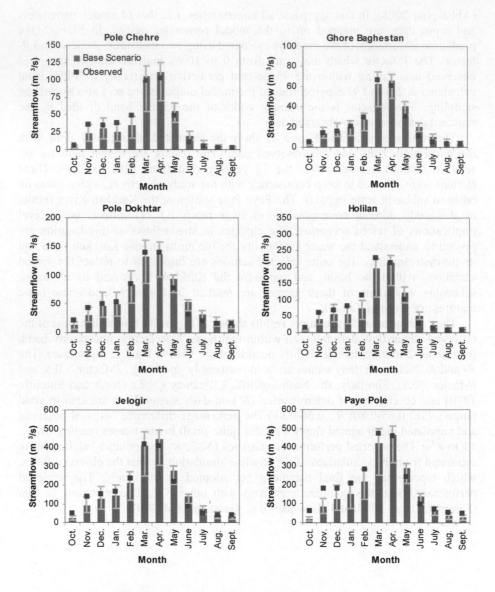

Figure 37. Monthly summary of the calibration and uncertainty analysis results.
(The 95PPU band is shown by thin green bars)

7.2.2. Tested scenarios

Three scenarios related to increased water consumption in the rain-fed agriculture were simulated and their impacts on streamflows evaluated against the baseline simulation. The changes in the mean annual and mean monthly streamflows were the main assessment indicators.

Scenario 1 (S1): Upgrading rain-fed areas to irrigated agriculture. In this scenario, the impact of upgrading rain-fed farming to irrigated agriculture was investigated. The rain-fed lands located in the valleys close to rivers with soils favorable for agriculture (e.g., alluvial soils) were considered as potential areas for irrigation. The GIS based overlay analysis conducted through SWAT interface was used to estimate the potential areas. The analysis revealed that a total area of about 0.5 million ha could be potentially upgraded from rain-fed to irrigated agriculture in the study basin (Table 24). This accounts for about 11% of the total study area. It is recognized that exact estimation of total irrigable area might vary, depending upon various physiographic, chemical, hydrological, topographic, social and economic factors. Moreover, we also consider that it will not be possible to convert all the rain-fed systems to irrigated ones due to several reasons as aforementioned. We assume that investments in developing surface water use through gravity based systems, lift irrigation schemes, direct pumping from the rivers and building small tanks and dams for small scale irrigation could contribute to the proposed land use shift (upgradation of rain-fed systems to irrigated ones) in future. Impact of such a shift on downstream water availability is not well known and will be investigated in this scenario.

S1 was represented in SWAT by using its water use routine, which provides options to specify average daily water consumption rates for each month. The water use can be defined from every subbasin through four possible sources of water, i.e., rivers, shallow aquifers, deep aquifers and ponds. We specified water consumption from the rivers only. The other options were not tested due to limitations mainly related to data availability.

The average daily water consumption from each of the 71 river reaches was defined based on the irrigation water demand from the potential rain-fed area considered for upgradation to irrigation. The water consumption was estimated for every month using the following equation.

$$ IWC = A_{ir}\left(E_{pot} - E_{act}\right) \qquad (18) $$

where, *IWC* is the average monthly irrigation water consumption rate in m^3/d, A_{ir} is the area upgraded to irrigation in m^2 and E_{pot} and E_{act} are the average monthly potential and actual evapotranspiration rates expressed in m/d. The E_{pot} was estimated using Hargreaves method (Hargreaves et al., 1985), whereas the actual evapotranspiration (E_{act}) was estimated based on the available moisture in the soil profile and evapotranspiration demand from rain-fed systems under wheat cultivation, as per SWAT water balance calculations under the baseline scenario. The A_{ir} was estimated by using information on land use, soil and slope data as mentioned above. The GIS based overlay analysis conducted through SWAT interface was used to estimate A_{ir}. The analysis revealed that a total area of about 5,000 km^2 (0.5 million ha) could be potentially upgraded from rain-fed to irrigated agriculture in the study area (Table 24).

The monthly estimates of E_{pot}, E_{act}, and difference between them used in the formulation of $S1$ are presented in Figure 38. On the whole, the required IWC was estimated around 275 mm for the period November to June, which generally reflects the growth season of winter wheat. Wheat is the main crop cultivated in the study area. The other crops include barley, chickpea, alfa alfa, maize, sugarbeat and vegetables. Wheat is cultivated from November to June. Maize and sugerbeat are cultivated during June to October, and are mainly grown in regions where irrigation supplies are available. Fodder (alfa alfa) and vegetables are grown in both cold and warm seasons. The calculations of irrigation requirements were constrained to a single cropping season reflecting wheat growth period (winter-spring). A double cropping system was not considered mainly due to the limitations of surface water supplies during summer season (Masih et al. 2009).

Table 24. The extent of area under drainage, rain-fed systems and irrigable rain-fed systems at the basin and sub-basin levels.

River reach/ catchment ID	Name of river	Name of streamflow gauge	Drainage area (DA) (million ha)	Area under rain-fed systems (million ha)		Rain-fed area convertible to irrigated area, (million ha)	
				million ha	% of DA	million ha	% of DA
20	Gamasiab	Pole Chehre	1.078	0.240	22	0.086	8
24	Qarasou	Ghore Baghestan	0.544	0.333	61	0.124	23
38	Saymareh	Holilan	2.042	0.736	36	0.253	12
60	Kashkan	Pole Dokhtar	0.952	0.282	30	0.099	10
66	Karkheh	Jelogir	3.952	1.399	35	0.468	12
71	Karkheh	Paye Pole	4.237	1.402	33	0.468	11

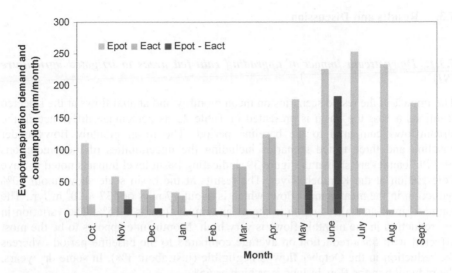

Figure 38. Values used in the development of S1 for the monthly potential evapotranspiration (E_{pot}), actual evapotranspiration (E_{act}), and the difference between E_{pot} and E_{act}.

Scenario 2 (S2): Improved soil water availability through rainwater harvesting.
Various in-situ water harvesting systems and soil and water conservation techniques are generally recommended as discussed in the introductory section of this chapter (e.g., micro-basins, terracing, bunds, and mulching etc.) to increase soil water retention and foster plant water availability. Such interventions can be represented in the SWAT model by modifying the Available Soil Water Capacity (AWC) parameter. The AWC parameter controls retention of water in the soil profile for consumption by plants. An increase in AWC generally leads to an increased soil water retention and thereby indirectly represents a soil and water conservation practice. Under S2, we assume that the recommended soil and water management interventions collectively increase AWC of the soils under rain-fed agriculture by 20%. A similar study was reported by Faramarzi et al. (2010), who investigated the impact of 20% increase in AWC on the irrigation requirement in Iran. However, their study did not evaluate the impacts on streamflows, and present study helps filling this important knowledge gap.

Scenario 3 (S3): Combined impact of S1 and S2. Under this scenario, the combined impact of the two scenarios (S1 and S2) was evaluated. S3 was represented in the SWAT model by keeping the water consumptions in the water use routine same as in the case of S1. Then the soil routine was modified by increasing the AWC parameter by 20% as done under S2. In this way, S3 simulated the combined effect of scenarios S1 and S2.

7.3.   Results and Discussion

7.3.1. Downstream impact of upgrading rain-fed areas to irrigated agriculture (S1)

The impact of the tested scenarios on mean monthly and annual flows at the selected locations across the basin is presented in Table 25 as percentage difference in the streamflows compared to the baseline period. The mean monthly flows under baseline and three tested scenarios including the uncertainties of the predictions (95PPU bands) are shown in Figure 39, indicating basin level impacts noted at Paye Pole station at the Karkheh River. The results at the basin scale show about 10% reduction in the mean annual flow, which is about 17 m^3/s or 537 × 10^6 m^3/yr.. The range of inter-annual reduction in the mean annual flow is 7-15%. The variation in the reduction in the monthly flows is very high. Month June appears to be the most affected with 56% reduction on average compared to the baseline period, whereas the reduction in the October flow is negligible (just about 1%). In some dry years, the reduction in the flow in June is as high as 65%.

The impacts vary notably among study subbasins (Table 25). These differences were mainly governed by the relative area brought under S1 and the amount of available flows. The highest impact was noted for Ghore Baghestan (Qarasou subbasin) where the highest proportion of area under S1 falls (Table 24). This subbasin indicated a decline of 15% in the mean annual flows. Monthly flow reductions were in the range of 0-92%. The inter-annual variation is also quite high, with annual flow reduction in the range of 10-43%. Monthly flow reductions could escalate further, reaching zero flow in June for some dry years. The Kashkan subbasin witnessed comparatively lower impact, where annual reductions were around 8%, varying from 0-52% between months, as shown by the estimates at Pole Dokhtar. The inter-annual variability was also modest here, with the annual flow reduction in the range of 6-11% and maximum monthly decline of around 66%.

In general, the highest flow reduction corresponds to June for all examined locations. The other months with high impacts are May, July, November and December. This pattern of monthly impact is somewhat similar across the basin. The considerable flow reductions were observed in July despite no water abstractions from streams during this month. This shows that a reduction in the streamflow in a month is likely to contribute into diminishing streamflow in the following month(s). This carryover impact is due to hydrological processes related to water storage in the river channel and subsequent contribution of stored water to the river flows. However, this impact was not prominent beyond July, as noted by the negligible change in August, September and October. Part of the irrigation abstractions generally contributes back to the streamflows in the form of return flows. A detailed investigation of the return flow processes was beyond the scope of this study, mainly due to the limitations related to the model and the available data.

Table 25. Difference in the mean monthly and annual streamflows, expressed in %, under the three tested scenarios as compared to baseline simulation.

Scenario/ Time level	Pole Chehre (Reach ID 20)	Ghore Baghestan (Reach ID 24)	Holilan (Reach 38)	Pole Dokhtar (Reach ID 60)	Jelogir (Reach ID 66)	Paye Pole (Reach ID 71)
S 1						
Oct.	0	0	0	0	0	1
Nov.	20	23	16	14	12	10
Dec.	9	15	14	6	12	12
Jan.	5	8	7	3	6	6
Feb.	3	5	4	2	3	3
Mar.	1	2	2	1	2	2
Apr.	1	3	2	1	2	2
May	21	31	23	13	18	17
June	89	92	77	53	61	56
July	21	22	37	10	30	38
Aug.	0	0	2	0	3	4
Sept.	0	0	0	0	1	2
Annual	10	15	12	8	10	10
S 2						
Oct.	2	3	3	1	2	2
Nov.	3	5	4	2	3	3
Dec.	3	6	4	3	4	4
Jan.	4	8	6	4	5	5
Feb.	6	9	7	3	6	5
Mar.	4	6	5	3	4	4
Apr.	2	4	3	2	3	3
May	2	3	2	2	3	3
June	1	3	3	2	3	3
July	1	3	3	2	3	3
Aug.	1	3	2	1	3	3
Sept.	1	3	2	1	2	2
Annual	3	5	4	2	4	4
S 3						
Oct.	2	3	3	1	2	3
Nov.	23	27	20	16	15	13
Dec.	12	21	17	9	15	15
Jan.	10	15	12	7	11	11
Feb.	9	14	11	5	8	8
Mar.	5	9	7	4	6	6
Apr.	3	7	5	3	5	5
May	23	33	25	15	20	19
June	89	93	77	54	61	56
July	22	25	39	11	32	40
Aug.	1	3	4	2	6	7
Sept.	1	3	2	1	3	4
Annual	13	20	16	10	14	14

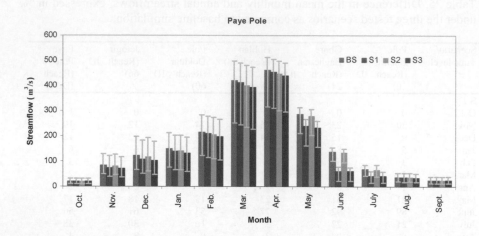

Figure 39.   Simulated streamflows for the baseline period and the three scenarios (S1, S2 and S3) at the basin level (Paye Pole station). Also shown in the figure are 95PPU uncertainty bands.

The results of S1 are instructive to guide the desired level of irrigation development by examining the corresponding spatio-temporal impacts. The expected reductions in the mean annual flows of 10% at the basin level and 8-15% across the subbasins seem quite reasonable. For instance, this would translate into an annual flow reduction of about 537×10^6 m³/yr. at the basin level. The water development potential of the basin was estimated to around $1\text{-}4 \times 10^9$ m³/yr., considering different levels of water allocations for the environment (Masih et al., 2009). This shows that the annual flow reduction at the basin level as a result of the S1 is well within the available water development potential of the basin.

However, the major concern in adopting S1 is related to excessively high percentage of flow reductions from May to July, most notably in June when reductions could be in the range of 50-100%. In general, the reductions exceeding 50% may severely impact downstream water needs. For instance, it would largely alter the natural flow regime of the river and is likely to have severe negative repercussions for the environment (e.g., Poff et al. 1997).

Thus, adoption of S1 would require additional considerations to avoid excessive decline in flows and consequent impacts on downstream uses and users. This could be achieved by reducing the abstractions, which could be done through decreasing the rain-fed area considered for upgradation. This option was further studied, and the results are substantiated by Figure 40, which shows the basin level impact on streamflows associated with a certain level of rain-fed area upgraded to irrigation for critical months of May, June and July. Such analysis could guide the choice of appropriate level of rain-fed agricultural development. In general, the results show that development of about one-fourth of the available potential rain-fed area to irrigation (e.g., about 0.1 million ha or 1,000 km^2) may be considered safe as it will keep the flow reduction in the most affected month June to below 30% ensuring reasonable levels of downstream water availability throughout the year.

The other complementary options for mitigating flow reductions may include: a) various forms of water storage to augment supplies during the most affected months (May and June), and b) practicing supplementary irrigation. The studies conducted by van der Zaag and Gupta (2008), and McCartney and Smakthin (2010) discussed a number of water storage options and pointed out that such options are also likely to address the issues of high variability in water availability due to observed and predicted climate variability and change. A considerable benefit of supplementary irrigation in terms of improving rain-fed agriculture in semi-arid to arid environments was shown by Oweis and Hachum (2009).

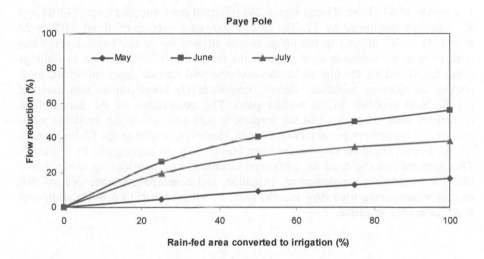

Figure 40. Impact on monthly streamflows (May-July) due to proportion of area upgraded from rain-fed to irrigated agriculture at the basin level.

7.3.2. Downstream impact of improved soil water availability through rainwater harvesting Scenario (S2)

Under S2, a reduction of about 4% or 6 m^3/s (194 × 10^6 m^3/yr.) was observed in the mean annual streamflow at the basin level (Table 25 and Figure 39). The impact at monthly time resolution was also small (2 to 5% decline) at the basin level. At the subbasin level, the annual flow reduction was in range of only 2 to 5%. Monthly flow reductions vary from 1-9 % across the basin. However, reductions in February-April were slightly higher in case of S2 than S1. This difference could be attributed to change in precipitation partitioning processes under S2, whereby, enhanced water retention in the soil and increased evapotranspiration diminish surface and subsurface runoff. This affect appeared to have more impact on streamflows than that of water diverted from rivers for irrigation during these months. The inter-annual pattern of changes indicated a decline in the range of 2-10 % and 0-20 % at annual and monthly time scales, respectively. The results suggested that the expected reductions in streamflows are reasonably small compared to the amount of water available for further development (as noted in the previous section). Therefore, it could be recommended to increase attention for the promotion of improved soil and water conservation practices in the basin.

7.3.3. Combined impact of S1 and S2 (S3)

The results of S3 (Table 25 and Figure 39) reflected the combined impact of S1 and S2, although dominated by S1. The results showed a decline of about 14% or 23 m^3/s (718 × 10^6 m^3/yr.) in the mean annual streamflow at the basin level. Flow reductions at the subbasin level were in the range of 10-20%. Similar to findings noted for S1 and S2, the highest impact was observed for two upper subbasins, most notably the Qarasou subbasin, whereas, comparatively lower impact was found in the Kashkan subbasin in the middle parts. The observation of the annual flow reductions under S3 shows that the scenario is still well within the available water resources development potential in the basin. However, similar to the findings under S1, the impact could be very severe from May to July, as highlighted by Table 25. This substantiates the need for additional attention, e.g., considering reduced area under S1, practicing supplementary irrigation, and developing storage options that can tap water during high flow months (e.g., February-April) to provide additional supplies in May and June.

7.3.4. Consideration of prediction uncertainty of the model

It is well recognized in hydrology that there are uncertainties involved in hydrological modeling arising from limitations of model structure, input data and parameterization (e.g., Beven 2001). These uncertainties not only influence the model calibration process but could also impact the predictions made by the model. For this study, a reasonable effort was designated to access the model uncertainty as per used procedure of SUFI-2. The reasonably good values of the uncertainty descriptors (e.g., P- factor>0.5 and R-factor<0.5) were obtained beside good results on commonly used performance measures in hydrological modeling (e.g., *NSE: 0.65-0.80, R^2 : 0.66-0.81* and volume balance ranging from -13 to 8%). Despite reasonably good values of the reported performance measures, the range of 95PPU band obtained from the final 500 parameter sets is still not small. Therefore, the choice of the parameter set used for simulating the baseline case may have implications for the studied impact scenarios. In this study, we adopted most commonly used approach of selecting a parameter set for baseline that produces simulations close to observed streamflows and results in reasonably good values of studied performance indicators (NSE, R^2 and volume balance). Therefore, the impact results of tested scenario reported in the previous sections were considered appropriate to guide decision-making process. However, the outcome should be considered cautiously, i.e., particularly considering the associated uncertainty in the results.

Furthermore, to get an overview of the prediction uncertainty, the range of prediction uncertainty associated with the simulated scenarios was estimated by comparing the values of upper and lower 95PPU band achieved under baseline case with the corresponding values of the tested scenarios. Figure 41 shows the percentage reduction in the mean annual streamflows along with the expected uncertainty. The results for S1 showed that the reduction of 10% in the mean annual streamflow at the basin level could have uncertainty range of 8-16%. Monthly results are shown in Figure 42, which, for instance, indicated that the uncertainty range of 50-77% could be associated with the values of flow reduction of 56% in June at the basin level. Furthermore, uncertainties of the different scenarios (e.g., S1 vs S2) and that of their flow reductions are markedly different. The uncertainty of S1 is dominating S3 uncertainty as well.

The major implications of the inclusion of prediction uncertainty results in impact evaluation and consequently consideration in the decision-making process could be: a) The range of flow reduction at annual scale including the uncertainty band still remains well within the available development potential in the study basin, e.g., 8-16% reduction at the annual scale at the basin level under S1., b) The range of monthly flow reductions including uncertainty clearly reveal much higher likelihood of flow reduction, e.g., the expected flow reduction in June could be in the range of 50-77% at the basin scale under S1. Therefore, considering the high range of uncertainty in the predicted impacts on monthly flows under S1, it could be stressed that the decisions regarding rain-fed area development would require additional considerations so that streamflows may not be excessively depleted during critical months (e.g., May-July)., and c) The impact of including prediction uncertainty in the case of S2 is small, which further strengthen the argument about promoting soil and water conservation techniques in the study basin.

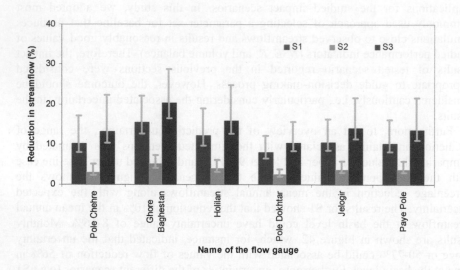

Figure 41. Assessment of the prediction uncertainty of the modeling results at annual time resolution under the three tested scenarios.

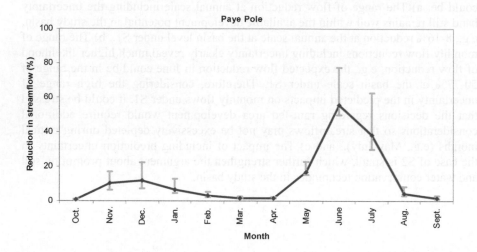

Figure 42. Assessment of the prediction uncertainty of the modeling results at monthly time resolution at the basin level.

7.4.   Summary and Concluding Remarks

The upgradation of the rain-fed systems through improving soil and water conservation practices and providing water for irrigation is of critical importance for the global food security, particularly for the rural poor living in water-scarce semi-arid to arid regions. However, a proper understanding of the benefits harnessed by developing water in upstream areas and the consequent impacts on downstream regions are important for informed planning and sustainable management of natural resources. This study contributes to such understanding by evaluating the impact of three scenarios of upgrading rain-fed areas using SWAT model. The tested scenarios were: upgrading rain-fed areas to irrigated agriculture (S1), improving soil water availability through rainwater harvesting (S2), and a combination of S1 and S2 (S3). The impacts on monthly and annual streamflows were investigated.

The basin scale impact of the tested scenarios suggested a decline in mean annual flows of about 17 m^3/s (537 × 10^6 $m^3/yr.$), 6 m^3/s (194 × 10^6 $m^3/yr.$) and m^3/s 23 (718 × 10^6 $m^3/yr.$) under scenarios S1, S2 and S3, respectively, when compared with the baseline case. This would mean a reduction of about 10%, 4% and 14% in case of S1, S2 and S3, respectively. The results revealed that the conversion of rain-fed areas to irrigation (S1) would have comparatively higher reductions in the downstream flows as compared to conserving water for plant uptake in the soil root zone through rain water harvesting (S2). In general, the estimated reductions in the mean annual streamflows at the basin scale fall well within the limits of available estimates of water resources development potential. However, under S1, monthly flows would be severely reduced from May to July, with highest impact in June when flows could reduce more than half of the available flows at the basin scale. This situation is much more alarming for two upstream subbasins (Gamasiab and Qaraou) where reduction in June flows could reach over 90 %. The noted high levels of impact on streamflows suggested the need of additional measures to avoid excessive proportions of flow reduction in these months. The excessive impacts could be minimized by reducing the area brought under irrigation (particularly in upper parts of the basin), developing a range of storage options to augment supplies and supplying less water than actually required (practicing supplementary irrigation).

The consideration of model prediction uncertainty reveal that the range of annual flow reduction is not large under all the tested scenarios (e.g., only 8-16% at the basin scale under S1) and, thus, model uncertainty is less likely to have any major implications in decision-making process in the context of annual flows. However, the range of monthly flow reductions was quite large when considering model prediction uncertainty, particularly for May-July, which further substantiate the need of adopting policy options to mitigate excessive flow reductions during these months.

Based on the results of this study, it could be recommended that the upgrading rain-fed areas should be concentrated in the middle parts of the basin (e.g., Kashkan River subbasin), through introducing irrigation and range of water storage options as well as promoting soil and water conservation techniques. The water management interventions in the upper subbasin areas of Gamasiab and Qarasou Rivers should mainly limit to promoting soil and water conservation techniques. In these areas, the conversion of potential rain-fed lands to irrigated ones should be promoted very cautiously with much attention given to additional considerations of developing fewer amounts of rain-fed areas, providing means of water storage to augment supplies and practicing supplementary irrigation.

In general, the methods used and the findings of this study are instructive for the other basins, and have demonstrated the importance of a rigorous spatio-temporal investigation of impacts of agricultural water management interventions across a large river basin.

8.   SYNTHESIS, CONCLUSIONS AND RECOMMENDATIONS

8.1.   Nature and Causes of a High Level of Hydrological Variability

A comprehensive analysis of daily streamflow records available over the period 1961-2001 at the seven very important flow gauging stations across the Karkheh River and its major tributaries revealed that the streamflows have large intra-annual and inter-annual variability across the basin. The high flows are observed from November to May, with peak flows occurring in March-April. The high flood events (1-day maximum) can occur anytime from November to April, though most often they occur in February and March. The low flow period corresponds to June through October. There are large differences between the amount of water available during high and low flow periods. For example, at the Jelogir station at the Karkheh River, mean monthly streamflow in April (386 m^3/s) is nearly ten times higher than in September (41 m^3/s). The observed spatio-temporal variations could be substantiated by the values of Coefficient of Variation (CV). Monthly CV values range from 0.4 to 1.77 across the examined streamflow gauges. In temporal terms, the minimum and maximum CV values correspond to February and November, respectively, whereas, in spatial terms, the Gamasiab River indicated higher variability and the Kashkan River the lowest variability. The mean annual streamflow indicated CV values in the range of 0.41 to 0.54, indicating marked differences in water available in the long-term perspective. For example, the mean and median surface water availability at the Paye Pole station at the Karkheh River was estimated as 5.83×10^9 m^3/yr. and 5.59×10^9 m^3/yr. As in all other examined stations, the minimum and maximum had a wide range at Paye Pole, with values of 1.916×10^9 m^3/yr. observed during 1999-2000 and 12.60×10^9 m^3/yr. observed during 1968-69. Under such highly variable conditions, the understanding of the reliability of the water availability becomes more meaningful for better resources use and allocation planning. The flow duration analysis conducted in this study provides such estimates of streamflow reliability for the Karkheh Basin at daily, monthly and annual time resolutions. For instance, the value of mean annual streamflow with a reliability level of 75% (indicated by 75th percentile of streamflow derived from the flow duration analysis) at Paye Pole was 4.10×10^9 m^3/yr., which is about 30% lower than the mean annual flow estimated for this location.

High climate variability is considered as the major driving factor of the observed spatio-temporal variability of the streamflows, among other factors such as soil, land use and geological characteristics. The streamflow regime also depicts notable differences with regard to spatial location in the study basin. For instance, more runoff is generated from the middle parts of the basin (e.g., Kashkan River catchment) compared to the upper parts of the basin (e.g., Gamasiab and Qarasou River catchments). These differences were mainly attributed to higher precipitation rates and lower water use by the agriculture sector in the catchment areas of the Kashkan River compared to the two upper catchments. Furthermore, the contribution of base flow in the total streamflow is higher for the Kashkan River compared to the other two upper catchments (Gamasiab and Qarasou), which could be due to the differences in land use, soil and geological characteristics, and therefore warrant further research.

8.2.   Water Allocations, Water Availability and Sustainability

The study reveals that the Karkheh Basin still appears to be an open basin, indicating some room for further water resources development. The estimated range of further water resources allocations was $1\text{-}4 \times 10^9$ m^3/yr., depending on the amount of water left for environmental flows. However, the water allocations should be done after a careful impact assessment and trade-off analysis for multiple and highly competing uses and users across the basin, which warrant a proper impact assessment before implementation. The review of ongoing water allocation planning shows that the allocation to different sectors of water use will be 8.90×10^9 m^3/yr. by the year 2025, among which the irrigation share will be the biggest (7.42×10^9 m^3/yr.), almost equaling the renewable water supplies available in an average year. Therefore, considering the water availability and its variability and water resources development plans, it is evident that the basin will very likely approach closure stage during the first quarter of this century. Meeting the demands of all users (particularly agriculture, hydropower and environment) will then be an extremely challenging task, particularly during dry years.

8.3.   Streamflow Trends and Their Underlying Causes

The investigation of trends in the hydro-climatic variables revealed a number of significant trends, both increasing and decreasing. The observed changes were nonuniform in term of their spatio-temporal prevalence. The upper parts of the basin, particularly Qarasou River, faced a notable decline in the low flow regime. A significant decline in the streamflow was observed in most of the studied low flow indicators, i.e., May, August, 1 and 7 days minima, low pulse count and duration for Ghore Baghestan station at the Qarasou River. On the other hand, the flood regime and winter flows indicated intensification in the middle parts of the basin, indicated by significant upward trends observed in 1 and 7 days maxima, high pulse count, October and December flows for Pole Dokhtar station at the Kashkan River. Further, downstream propagation of the observed trends was found dependent on the combined effect of the upstream drainage areas. For instance, the declining pattern during low flow months was not significant for the Karkheh River, because the declining trends that emerged from the upper catchments (Gamasiab and Qarasou) were counterweighted by the stable low flow behavior of the middle catchments (e.g., Kashkan). However, the significant trends observed in a number of streamflow variables at Jelogir, 1-day maximum, December flow, and low pulse count and durations indicated changing hydrological regime of the Karkheh River. Most of the observed trends were found triggered by the changing climatic behavior, observed at a number of studied synoptic climatic stations. The study suggests that the decline in April and May precipitation caused the decline in the low flows while the increase in winter (particularly March) precipitation coupled with temperature changes led to an increase in the flood regime. The catchment degradation could be a complementary factor in the intensification of the flood regime, while increased water abstractions might have additionally contributed in the declining low flow regime.

8.4. Addressing Methodological and Data Scarcity Issues in the Hydrological Modeling

The use of hydrological models is generally seen pivotal in better understanding the hydrological processes and has become a norm to test various "what if" scenarios, which otherwise could not be well investigated on the basis of observed data alone. This study reveals that data scarcity remains a major challenge in the basin-wide hydrological modeling and water resources assessment, but could be addressed through developing innovative and tailor-made methodologies and innovative solutions for a study basin.

The Karkheh Basin noted abandoning of various flow gauging stations during the course of time. Estimation of streamflow records for these poorly gauged catchments emerged as an important issue in the study area, and is also generally seen as a major challenge in hydrology. A new regionalization method was developed in this study. The proposed method is based on the regionalization of a conceptual rainfall-runoff model (the HBV model) parameters whereby model parameters could be transferred from gauged catchment to the poorly gauged catchment depicting hydrological similarity defined, based on the similarity in their flow duration curves (FDC). It was demonstrated that the FDC-based regionalization method worked well in the data limited Karkheh Basin as compared to other widely recommended methods (e.g., spatial proximity and catchment similarity defined from the physiographic and climatic characteristics). Moreover, the new FDC-based regionalization method is regarded as a good addition to the available regionalization method, as it compared very well with most of the available methods tested in other countries.

Better representation of precipitation data in the hydrological modeling also emerged as an important consideration in the hydrological modeling of the Karkheh Basin, besides its general recognition and research needs in hydrological modeling. It is well recognized that the climatic data are the major driver of the hydrological and other processes simulated by a model. In this regard, the benefit of using areal precipitation derived from the observed station records by using inverse distance and elevation weighting method was evaluated against the usual way of using station data in the semi-distributed SWAT model. The study showed that the SWAT streamflow simulations improved for smaller catchments (600-1,600 km^2), most of them having poor density and distribution of rain gauges, whereas larger catchments (>5,000 km^2) could be modeled equally well under both cases, mainly attributed to the averaging-out effect of precipitation at larger catchment-to-basin scales. The study demonstrated that the use of areal precipitation improved simulated streamflows; however, the results were influenced by the spatial scale of the investigation and distribution and density of rain gauges.

The gained understanding of the hydro-climatic variables and processes, i.e., through system investigation, innovative ways of improving available but scarce streamflow and precipitation data, field visits and discussions with the stakeholders, and a literature review, appeared to be instrumental in the good calibration of the SWAT model for the upper mountainous parts of the Karkheh Basin (42,620 km^2) from where almost all of the basin's runoff is generated. Moreover, better understanding of the observed processes and improved quality of the input data together with rigorous exercise of parameter estimation (based on both manual and automatic procedures) and uncertainty analysis helped reduce the prediction uncertainty of the used model, thereby providing a reasonably good confidence in the modeled hydrological processes and investigation of various "what if" scenarios.

8.5.   Consideration of the Impacts on Downstream Water Availability while Upgrading Rain-fed Agriculture

The well-calibrated SWAT model was used to test the impacts of soil and water management interventions in the rain-fed systems in the upstream areas on the streamflows in the downstream areas. The impacts of upgrading rain-fed systems to irrigated agriculture (S1), soil and water conservation practices (S2), and a combination of S1 and S2 (S3) were studied on the mean annual and monthly streamflows at catchment to basin levels. The results reveal that the expected reduction in the mean annual flows (e.g., about 10% under S1, 4% in S2 and 14% in S3) remains well within the available development potential in the basin, even after consideration of the model's prediction uncertainty. However, excessive decline in flows during May-July (in particular during June) warrants additional measures to ensure downstream water availability and environmental integrity throughout the year. These excessive impacts could be minimized by reducing the area brought under irrigation (particularly in upper parts of the basin), supplying less water than actually required (practicing supplementary irrigation), and developing a range of storage options to augment supplies.

8.6.   Contribution and Innovative Aspects of This Research

This PhD research addresses some key issues related to basin wide spatio-temporal assessment of hydrology and water resources and highlights its importance and use in water resources planning and management in the river-basin context. An attempt has been made to specifically address various issues. The relevance of the study objectives and research questions to the international literature, specific contribution to the scientific debate, importance with respect to the case study basin, and innovative aspects are highlighted in each of the chapter on results and discussions (e.g., Chapter 3-7). The main innovative aspects and contributions of this research include, but are not limited to: (i) development of a new FDC based regionalization approach, (ii) innovative use of areal precipitation input and evaluation of its implications on streamflow simulations in a large basin using the SWAT model, (iii) contribution to improve understanding of the streamflow trends and their linkages with climate, and (iv) improved knowledge on spatio-temporal variability of hydrology and water resources, through use of rigorous state-of-the-art methods, including development and application of new innovative techniques.

8.7. Major Recommendations and Future Directions

Based on the study findings, the following major policy actions are recommended. The ongoing water allocation planning is not sustainable and its thorough revision is recommended. The sectoral water allocation needs to be revised in the light of resource availability and variability, a sound foundation of which has been laid in this study. In view of the high share of water allocation for human demands (particularly agriculture), the environment is likely to suffer the most in the near future. Therefore, further assessments of the environmental water needs for in-stream, floodplain and Hoor-Al-Azim Swamp are highly recommended. Although a detailed assessment of environmental flow requirements was beyond the scope of this study, the preliminary estimates could be based from the hydrological assessments carried out in this study (see chapters 3 and 4), before more detailed information on environmental needs and those of other sectors become available.

The changing climate and hydrological regime in the basin further added to the complexity of hydrological and water management issues and require immediate attention. Considering the nonuniform nature of the observed trends, the adaptation response should be underpinned by concurrent nonuniform but basin-wide approaches that include a sound understanding of spatio-temporal differences in the observed trends as well as their interactions. For instance, the declining low flows (e.g., May through September) in the upper parts of the basin (Qarasou and Gamasiab subbasins) could be tackled through various strategies, i.e., introducing restriction on the use of surface water during low flow months in these areas. Moreover, the mitigation of an intensified flood regime, particularly in the middle parts of the basin (e.g., Kashkan subbasin) should receive high priority in the short term, to avoid negative repercussions on life, infrastructure and socioeconomics in the Karkheh Basin. This should also remain a major policy focus in the long-term strategy, as the predicted climate change is expected to increase frequency and magnitude of floods in the study region (e.g., Abbaspour et al. 2009). Efforts should also be enhanced to reduce the degradation of land cover (particularly forest and rangelands in the middle parts of the basin), as the improved land cover is likely to help stabilize the runoff response.

In situ soil and water conservation techniques should be promoted across the rain-fed systems, as they are likely to pose a small impact on the downstream water availability. However, the upgradation of rain-fed systems to full-scale irrigated agriculture should be carried out partly and very cautiously to avoid jeopardizing the downstream demands from the environment, hydropower, irrigation and other uses during low flow months in particular.

Although the study has demonstrated that various innovative solutions could help cover the data gaps, more studies and investments should be made on data collection and better use of available (scarce) data sets. The abandoning hydro-climatic monitoring network across the Karkheh River system should be overhauled. Recently, more climatic stations have been installed in small cities, but the coverage generally remains poor for the mountainous parts, undermining proper hydrological investigations.

The SWAT model application demonstrated in this study should be extended, i.e., by including sedimentation and water-quality processes, and by testing other "what if" scenarios (e.g., related to storage options, land use changes, and environmental flows).

In general, the study provided a scientifically important and practically relevant example of hydrological assessment and its use in the water resources planning and management in the river basin context, which is instructive for the Karkheh and other river basins of Iran, and worldwide.

SAMENVATTING[10]

De escalerende toename van watergebruik voor menselijke doeleinden, met name voor landbouw, leidt tot toenemende druk op de zoetwatervoorraden. Alhoewel de toe-eigening van het water de mensheid op vele manieren heeft geholpen zoals het verbeteren van de voedselvoorziening en het social-economisch welzijn, heeft het ook tot schade geleid aan het milieu en de daaraan gerelateerde voorzieningen. Het evenwicht tussen mens en natuur met betrekking tot watergebruik wordt gezien als de grootste uitdaging van deze eeuw. Dit is nog veel ingewikkelder voor de semi-aride tot aride gebieden in de wereld, zoals de Islamitische Republiek Iran, waar water over het algemeen schaars is en de vraag vanuit de landbouw, industrie, verstedelijking en de groeiende bevolking snel toenemen. Door de aanwezige variatie in klimaat en de verwachte klimaatveranderingen zullen de problemen alleen nog maar toenemen.

In het geval van waterschaarste en concurrerend watergebruik, is een betere kennis van de stroomgebiedshydrologie en waterbeschikbaarheid essentieel voor beleidsvorming en duurzame ontwikkeling binnen de watersector. Deze studie is uitgevoerd in het semi-aride tot aride Karkheh stroomgebied in Iran, waar grootschalige waterverdeelplannen voor handen zijn, maar waar een uitgebreide kennis van de stroomgebiedshydrologie en het effect van de waterbeheersplannen op het watergebruik en de watergebruikers in het stroomgebied ontbreekt. De belangrijkste doelstelling van dit onderzoek is om een hydrologische studie van de (oppervlakte) waterbeschikbaarheid van het Karkheh stroomgebied te maken en het omvang van de variatie en de verandering te bestuderen op verschillende tijds- en plaatsafhanklijke schalen. De gebruikte methode bestaat uit een gecombineerd gebruik van een gedegen systeem analyse en hydrologische modeleer technieken. De plaatsafhankelijke dimensie is bestudeerd op stroomgebied, deelstroomgebied en substroomgebied, terwijl de tijdsafhankelijke dimensie is bestudeerd aan de hand van dag, maand, jaarwaarden en lange tijdsreeksen.

De uitgebreide hydrologische studie van de plaats- en tijdsafhankelijke variatie in het oppervlaktewater is uitgevoerd op basis van een lange tijdsreeks van dagelijkse afvoergegevens tussen 1961 en 2001 voor zeven belangrijke afvoermeetstations gelegen in de Karkheh rivier en haar belangrijkste zijtakken. De analyses zijn uitgevoerd met behulp van verschillende technieken, waaronder centrale tendentie en dispersie, afvoerscheiding en debietduuranalyse. Tevens is stroomgebied boekhouding toegepast voor het jaar 1993-94, waarvoor de benodige gegevens beschikbaar waren.

Het onderzoek laat zien dat de hydrologie van het Karkheh stroomgebied grote inter- en intra-jaarlijkse variatie heeft, voornamelijk veroorzaakt door grote plaats-

---

[10] Summary in Dutch was translated by Susan Graas and Marloes Mul, UNESCO-IHE, Delft, the Netherlands.

en tijdsafhankelijke variatie in klimaat en lokale verschillen in bodem, landgebruik en hydrogeologische eigenschappen van het stroomgebied welke grotendeels onderdeel uitmaakt van het Zagros gebergte. De afvoer neemt toe vanaf begin oktober en duurt tot aan april. Piekafvoeren vinden normaliter plaats in maart en april, maar overstromingen kunnen te allen tijden tussen november en april plaatsvinden. Deze hoge afvoeren worden veroorzaakt door een combinatie van smeltwater en neerslag. In de periode mei tot en met september overheersen lage afvoeren afkomstig van de basis afvoer vanuit de ondiepe grondwatervoorraden. Opvallend is dat het afvoerregiem in het midden gedeelte van het stroomgebied (Karkheh rivier) verschilt van de bovenstroomse gedeelten (Gamasiab en Qarasou), waarbij de eerste meer afvoer per oppervlakte-eenheid genereert en een hogere basisafvoer heeft. De kwestie van variatie is hier onderbouwd middels ramingen van de gemiddelde jaarafvoeren van de Karkheh rivier gemeten bij het Paye Pole station (net benedenstrooms van de Karkheh dam). De gemiddelde afvoer op deze locatie is $5,83 \times 10^9$ m³/jaar, terwijl de jaarafvoer in het extreem droge jaar 1999-2000 slechts iets meer dan eenderde ervan bedroeg ($1,916 \times 10^9$ m³/jaar) en het hoogst gemeten jaarafvoer was gemeten gedurende het extreem natte jaar 1968-69 en bedroeg circa $12,60 \times 10^9$ m³/jaar. Onder dergelijk sterk varierende omstandigheden, is het begrip van de betrouwbaarheid van informatie met betrekking tot waterbeschikbaarheid zeer belangrijk voor beter watergebruik en beslissingen omtrend watertoewijzing. Voor het Karkheh stroomgebied is de infornatie over de variatie in afvoer en betrouwbaarheid op dag, maand en jaarbasis verkregen door middel van de debietsduuranalyse.

De synthese van de resultaten over de hydrologische variatie, waterbeschikbaarheid, en water boekhouding suggereert dat het Karkheh stroomgebied open was gedurende de onderzoeksperiode (1961-2001), en dat er ruimte is voor verdere watertoewijzing, tot circa $1\text{-}4 \times 10^9$ m³/jaar, afhankelijk van de hoeveelheid water toegewezen aan de natuur. Echter, de toewijzing dient pas te gebeuren na een zorgvuldige effectenstudie en trade-off analyse van meerdere en zeer concurrende toepassingen en gebuikers in het stroomgebied. De op hande zijnde waterbeheerplannen lijken niet duurzaam te zijn gelet op de beschikbare hoeveelheid water en zijn grote variatie. Indien het huidige waterbeleid ook in de toekomst zal worden uitgevoerd dan zal het stroomgebied uiterlijk in 2025 gesloten zijn en dan zal tegemoetkomen aan alle watervragen zeer moeilijk zijn, met name gedurende de maanden met lage afvoer en tijdens droge jaren. Het milieu zal waarschijnlijk het meeste schade ondervinden aangezien deze, tot op heden, de laagste prioirteit heeft gekregen, maar ook andere sectoren waaronder de landbouw en huishoudelijk gebruik zullen waarschijnlijk ook te maken krijgen met vermindering van hun toegewezen waterrechten.

Als onderdeel van de systeemanalyse zijn tevens de veranderingen in de hydro-klimatologische variabelen en hun afhankelijkheid onderzocht. Afvoergegevens van vijf meetstations zijn gebruikt voor de periode 1961-2001 om trends in een aantal afvoer variabelen te onderzoeken welke de omvang van afvoervariatie laten zien, bv gemiddelde jaar en maandafvoer, 1- en 7-daagse maximum en minimum afvoer, datum van de 1-daagse maxima en minima en het aantal en duur van piek- en lage afvoeren. Voor de neerslag- en temperatuurgegevens van zes synoptische klimaatstations in de periode 1950 tot 2003 is een vergelijkbaar onderzoek naar trends in klimatologische variabelen uitgevoerd alsmede de correlatie met de afvoer. De Spearman rank test is gebruikt voor het vaststellen van de trends en de correlatie analyse is gebaseerd op de Pearson methode. De resultaten laten een aantal significante trends in afvoervariabelen zien, zowel toenemend als afnemend. Bovendien zijn de gevonden trends niet uniform qua plaats. De afname in de basisafvoer karakteristieken zijn significanter in de bovenstroomse delen van het stroomgebied (met name de Qarasou rivier), terwijl de toenemende trends in hoge afvoeren en winter afvoeren met name in het middengedeelte van het stroomgebied (Kashkan rivier) plaatsvinden. De meeste van deze trends worden voornamelijk veroorzaakt door veranderingen in neerslag. De resultaten laten zien dat de vermindering van neerslag in april en mei leidt tot vermindering in de basisafvoer, terwijl toename van de neerslag in de winter (met name in maart) samen met temperatuursveranderingen leidt tot een toename in het overstromingspatroon. De gevonden trends van het Jelogir meetstation aan de Karkheh rivier reflecteren het gecombineerde effect van het bovenstrooms gelegen stroomgebied. De gevonden significante trend voor een aantal afvoer karakteristieken waaronder het 1-daagse maximum, de december afvoer en het aantal en de duur van de lage afvoer, wijzen op veranderingen van het hydrologische regiem van de Karkheh rivier en worden voornamelijk toegewezen aan veranderingen in de klimatologische variabelen.

In het Karkheh stroomgebied zijn afvoergegevens van vele deelstroomgebieden niet beschikbaar zijn en veel afvoermeetstations zijn verlaten, hierdoor is de methode van het regionaliseren van de hydrologische parameters geschikt voor het bepalen van de waterbeschikbaarheid in die lokaties. In dit onderzoek is een nieuwe regionaliseringsmethode ontwikkeld om afvoertijdreeksen te schatten voor slecht bemeten stroomgebieden. De voorgestelde methode is gebaseerd op de regionalisering van een conceptueel neerslag-afvoer-model gebaseerd op de gelijkenis met de debietsduuranalyse (DDA). De resultaten van deze methode zijn vergeleken met drie andere methoden die gebaseerd zijno op de groote van het stroomgebied, de ruimtelijke nabijheid en de stroomgebiedskenmerken. De gegevens van 11 bemeten stroomgebieden (475 tot 2.522 km^2) zijn gebruikt om de regionaliseringsprocedures te ontwikkelen. Het op grote schaal toegepaste model HBV is gebruikt om dagelijkse afvoeren te simuleren middels overgedragen parameters van bemeten vergelijkbare stroomgebieden. Het onderzoek laat zien dat het baseren van de HBV parameters op de DDA vergelijkingsfactor een betere afvoer simuleert dan de drie andere methodes. Bovendien is aangetoond dat de onzekerheid van de parameters een klein effect heeft op het regionaliseringsresultaat. De resultaten van deze nieuwe methode verhouden zich zeer goed met de meeste elders ontwikkeld en toegepaste regionaliseringsmethoden. Daarom is de in deze studie ontwikkelde DDA regionaliseringsmethode een waardevolle aanvulling op bestaande regionaliseringsmethoden. De voorgestelde methode is eenvoudig te repliceren in andere stroomgebieden, met name die stroomgebieden waarvan het afvoerwaarnemingsnetwerk verminderd.

Hiernaast is een semi-gedistribueerd, proces-gebaseerd model – Soil Water Assessment Tool (SWAT) – gebruikt om de hydrologische fluxen te begrijpen en te kwantificeren en om verschillende scenarios te testen. Het is bekend dat het veel gebruikte SWAT model een groot verscheidenheid aan mogelijkheden biedt om de modelstructuur te definieren, maar de invoer van klimatologische gegevens is nog redelijk simplistisch. SWAT gebruikt de gegevens van het neerslagstation welke het dichst bij het zwaartepunt van dat deelstroomgebied gelegen is. Dit is wellicht niet representatief voor de neerslag in het hele deelstroomgebied en kan leiden tot toenemende onzekerheid in de model resultaten. In dit onderzoek is een alternatieve methode voor de invoer van de neerslag gegevens geevalueerd. Specifiek is de invoer van geinterpoleerde gebiedsneerslag getest versus de standaard SWAT invoer procedure voor neerslag. Het modelleergebied beslaat 42.620 km^2 en is gelegen in het bergachtige, semi-aride deel van het onderzoeksstroomgebied, welke het grootste deel van de stroomgebiedsafvoer genereert. De modelresultaten zijn beoordeeld op dag-, maand- en jaarbasis aan de hand van een aantal indicatoren van 15 afvoerstations, met een stroomgebied van 590 tot 42.620 km^2. De vergelijking suggereert dat het gebruik van gebiedsneerslag de modelprestaties verbetert, met name in kleine sub-stroomgebieden van 600 tot 1.600 km^2. De invoer van gebiedsneerslag resulteert in een toenemende betrouwbaarheid van gesimuleerde afvoer in gebieden met een kleine neerslagstations dichtheid en een slechte verdeling van de neerslagmeter(s). Beide methodes voor de invoer van neerslaggegevens resulteren in redelijk goede simulaties voor grotere stroomgebieden (meer dan 5.000 km^2), wat verklaard kan worden door het uitmiddelen van de neerslagvariatie voor grotere gebieden.

Het begrip van de stroomgebiedshydrologie aan de hand van de bovengenoemde studies, veldbezoeken en literatuuronderzoek en gedegen parameterschatting procedures heeft geholpen om een redelijk goede calibratie, validatie en onzekerheidsanalyse van het SWAT model voor het Karkheh stroomgebied te krijgen. Dit levert voldoende zekerheid op om het model toe te passen voor het analyseren van watergebruik scenarios in het stroomgebied. Drie scenarios, gerelateerd aan een toename van watergebruik in de regenafhankelijke landbouw, zijn geevalueerd. De onderzochte scenarios zijn: opwaarderen van regenafhankelijke gebieden naar geirrigeerde landbouw (S1), verbeteren van het bodemvochtgehalte door opvang van regenwater (S2) en een combinatie van S1 en S2 (S3). De resultaten van deze scenarios zijn vergeleken met de baseline in de periode 1988-2000. De baseline simulaties zijn uitgevoerd met de uiteindelijk vastgestelde modelstructuur en de parameters verkregen uit de gebruikte calibratie procedure. De resultaten van het eerste scenario (S1) geven een vermindering van 10% van de gemiddelde jaarafvoer op stroomgebiedsniveau, welke varieert van 8 tot 15% voor de belangrijkste deelgebieden van het stroomgebied. De afname in de gemiddelde maandafvoeren varieert tussen de 3 en 56% op stroomgebiedsniveau. De maanden mei – juli vertonen een groot effect, met in juni de grootste afname in afvoer. De afnames in afvoer in deze maanden zijn groter in de bovenstroomse gebieden van het stroomgebied wat hoofdzakelijk veroorzaakt wordt door een relatief groter potentieel te ontwikkelen irrigatiegebied in combinatie met relatief lagere afvoeren in deze maanden. De effecten van S2 zijn over het algemeen klein op deelgebied en stroomgebiedsniveau, met afnames van 2-5% en 1-10% in respectievelijk de gemiddelde jaar- en gemiddelde maandafvoeren. De geschatte afvoerafnames op jaarbasis blijft ruim binnen het beschikbare waterontwikkelingspotentieel van het stroomgebied. Echter, het voorkomen van buitensporige afvoerafnames in mei-juli zal aanvullende maatregelen vereisen, zoals aanvullende irrigatie en het vergroten van de aanvoer door een reeks van bergingsmogelijkheden en rekening houden met het opwaarderen van minder landbouwgrond naar irrigatie dan potentieel mogelijk is (met name in de bovenstroomse gebieden van het stroomgebied).

Het onderzoek concludeert dat kennis van de variatie in hydrologie en waterbeschikbaarheid, en het inachtnemen van de variatie van de waterbeschikbaarheid in de waterbeheerplannen een belangrijke rol speelt in het duurzaam gebruik and management van het beschikbare water in het Karkheh stroomgebied. De huidige watertoewijzing is niet duurzaam en een grondige herziening wordt aanbevolen. Deze zal uiteindelijk een vermindering in waterrechten voor menselijk gebruik (met name de landbouw) vereisen en leiden tot meer water voor het milieu. De klimatologische variatie en veranderingen hebben het afvoerregiem van de Karkheh rivier significant veranderd, wat onmiddelijke ingrijpen zou rechtvaardigen, door bijvoorbeeld structurele maatregelen en programma's om de stroomgebieddegradatie ten behoeve van het beheersen van overstromingen in het middengedeelte van het stroomgebied te herzien en het bekijken hoe wateronttrekkingen gedurende de lage afvoermaanden (mei tot september) in de bovenstroomse gebieden verminderd kunnen worden om de gevolgen van de afname in lage afvoeren in deze gebieden te compenseren. De uitgevoerde effectenstudie laat zien dat efficienter watergebruik in de neerslagafhankelijke landbouw gepromoot zou kunnen worden, met inachtneming van bodembeschermende en waterbesparende technieken in het hele stroomgebied aangezien deze minimale gevolgen hebben voor de waterbeschikbaarheid benedenstrooms. Echter, de omzetting van gebieden metgrotendeels neerslagafhankelijke landbouw naar irrigatie vereist een voorzichtige aanpak om redelijke limieten van afvoer afnames op maandbasis te verzekeren. Dit vereist het opwaarderen van beperkte gebieden met neerslagafhankelijke landbouw naar irrigatie (met name in het bovenstroomse gedeelte van het stroomgebied), het toepassen van irrigatie enkel bij tekorten en het ontwikkelen van een reeks aan waterbergingsmogelijkheden. Het versterken van hydro-klimatologische meetnetwerken wordt aanbevolen om de beschikbaarheid van gegevens en daarmee de toepassing van hydrologische en waterbeheermodellen voor beter geinformeerde besluitvorming te verbeteren. Hiermee samenhangend wordt het herstellen van verlaten hydro-klimatologische meetstations en het overwegen om meer meetstations in het berggebied te installeren aanbevolen. Het integraal waterbeheer dient gepromoot te worden in het onderzoeksgebied.

De kennis vergaard tijdens dit onderzoek kan als zeer relevant gezien worden voor andere stroomgebieden in Iran en wereldwijd.

REFERENCES

Abbaspour, K.C.; Johnson, C.A.; Genuchten, M.T. 2004. Estimating uncertain flow and transport parameters using a sequential uncertainty fitting procedure. *Vadose Zone Journal* 3(4): 1340-1352.

Abbaspour, K.C.; Yang, J.; Maximov, I.; Siber, R.; Bogner, K.; Mieleitner, J.; Zobrist, J.; Srinivasan, R. 2007. Modelling hydrology and water quality in the pre-Alpine/Alpine Thur Watershed using SWAT. *Journal of Hydrology* 333: 413-430.

Abbaspour, K.C. 2008. *SWAT-CUP2: SWAT calibration and uncertainty programs - A user manual.* Department of Systems Analysis, Integrated Assessment and Modelling (SIAM). Duebendorf, Switzerland: Eawag, Swiss Federal Institute of Aquatic Science and Technology.

Abbaspour, K.C.; Faramarzi, M.; Ghasemi, S.S.; Yang, H. 2009. Assessing the impact of climate change on water resources in Iran. *Water Resources Research* 45: W10434, doi:10.1029/2008WR007615, 2009

Abdulla, F.A.; Lettenmaier, D.P. 1997. Development of regional parameter estimation equations for a macro scale hydrologic model. *Journal of Hydrology* 197: 230-257.

Absalan, S.; Heydari, N.; Abbasi, F.; Farahani, H.; Siadat, H.; Oweis, T. 2007. Determination and evaluation of water productivity in the saline areas of lower Karkheh River Basin (KRB), Iran. In: *Extended abstracts. International workshop on improving water productivity and livelihood resilience in Karkheh River Basin,* Ghafouri, M. (ed.), Tehran, Iran: Soil Conservation and Watershed Management Research Institutes (SCWNRI), September 10–11, 2007.

Ahmad, M.D.; Islam, Md. A.; Masih, I.; Muthuwatta, L.P.; Karimi, P.; Turral, H. 2009. Mapping basin-level water productivity using remote sensing and secondary data in the Karkheh River Basin, Iran. *Water International* 34 (1): 119-133.

Ahmad, M.D.; Giordano, M. 2010. The Karkheh River basin: The food basket of Iran under pressure. *Water International* 35(5): 522-544.

Alcamo, J.; Henrichs, T.; Rosch, T. 2000. World water in 2025: Global modeling and scenario analysis. In: *World water scenarios analyses,* ed., Rijsberman Marseille: World Water Council.

Allen, R.G.; Pereira, L.S.; Raes, D.; Smith, M. 1998. *Crop evapotranspiration: Guidelines for computing crop water requirements.* FAO Irrigation and Drainage Paper No. 56. Rome, Italy: FAO, Water Resources, Development and Management Service.

Alijani, B. 2002. Variations of 500 hPa flow patterns over Iran and surrounding areas and their relationship with the climate of Iran. *Theoretical and Applied Climatology* 72: 41-54.

Alijani, B.; Ghohroudi, M.; Arabi, N. 2008. Developing a climate model for Iran using GIS. *Theoretical and Applied Climatology* 92: 103-112.

Ardakanian, R. 2005. Overview of water management in Iran. In: *Water conservation, reuse, and recycling, Proceeding of an Iranian American workshop,* pp. 153-172. Washington, DC: The National Academies Press.

Arnell, N.W. 1999. The effect of climate change on hydrological regimes in Europe: A continental perspective. *Global Environmental Change* 9: 5-33.

Arnold, J.G.; Srinivasan, R.; Muttiah, R.S.; Williams, J.R. 1998. Large area hydrologic modeling and assessment part 1: Model development. *Journal of the American Water Resources Association* 34(1): 73-89.

Arnold, J.G.; Allen, P.M. 1999. Validation of automated methods for estimating baseflow and groundwater recharge from stream flow records, *Journal of the American Water Resources Association* 35 (2): 411–424.

Arnold, J.G.; Fohrer, N. 2005. SWAT2000: Current capabilities and research opportunities in applied watershed modeling. *Hydrological Processes* 19: 563-572.

ASCE Task Committee on Definition of Criteria for Evaluation of Watershed Models of the Watershed Management Committee Irrigation and Drainage Division. 1993. Criteria for evaluation of watershed models. *Journal of Irrigation and Drainage Engineering* 119 (3): 429-442.

Ashrafi, S.; Qureshi, A.S.; Gichuki, F. 2004. Karkheh Basin profile: Strategic research for enhancing agricultural water productivity. Draft Report. Challenge Program on Water and Food. (Duplicated).

Balonishnikova, J.; Heal, K.; Guobin, F.; Karambiri, H.; Oki, T. 2006. World water resources, water use and management. Chapter 2, *Hydrology 2020: An integrating science to meet world water challenges,* ed. Oki, T.; Valeo, C.; Heal, K. IAHS Publication 300. Wallingford, UK: IAHS Press.

Beniston, M. 2003. Climate change in mountain regions: A review of possible impacts. *Climatic Change* 59:5-31.

Bergström, S. 1992. *The HBV model—Its structure and applications.* SMHI, no. 4. Norrköping, Sweden.

Beven, K.J. 2001. *Rainfall-runoff modelling: The primer.* Chichester: John Wiley & Sons.

Birsan, M.V.; Molnar, P.; Burlando, P.; Pfaundler, M. 2005. Streamflow trends in Switzerland. *Journal of Hydrology* 314:312-329.

Blöschl, G. 2006. Hydrologic synthesis: Across processes, places, and scales. *Water Resources Research* 42: W03S02, doi:10.1029/2005WR004319.

Blöschl, G.; Sivapalan, M. 1995. Scale issues in hydrological modelling: A review. *Hydrological Processes* 9 (3-4): 251-290.

Bouraoui, F.; Grizzetti, B.; Granlund, K.; Rekolainen, S.; Bidoglio, G. 2004. Impact of climate change on the water cycle and nutrient losses in a Finish catchment. *Climatic Change* 66:109-126.

Bouwer, H. 2000. Integrated water management: Emerging issues and challenges. *Agricultural Water Management* 45: 217-228

Burn, D.H.; Elnur, M.A.H. 2002. Detection of hydrologic trends and variability. *Journal of Hydrology* 255: 107-122.

Castellarin, A.; Galeati, G.; Brandimarte, L.; Montanari, A.; Brath, A. 2004. Regional flow–duration curves: Reliability for ungauged basins. *Advances in Water Resources* 27: 953-965.

CAWMA (Comprehensive Assessment of Water Management in Agriculture). (2007). *Water for food, water for life: A comprehensive assessment of water management in agriculture.* Colombo, Sri Lanka: International Water Management Institute (IWMI); London, UK: Earthscan.

Ceballos-Barbancho, A.; Morán-Tejeda, E.; Luengo-Ugidos, M.Á.; Llorente-Pinto, J.M. 2008. water resources and environmental change in a Mediterranean environment: The south west sector of the Duero River Basin (Spain). *Journal of Hydrology* 351: 126-138.

CPWF (Challenge Program on Water and Food). 2002. *Challenge program on water and food: Full proposal.* Challenge Program on Water and Food (CPWF). http://www.waterandfood.org/

CPWF. 2003. *Karkheh River Basin: Protecting dry land under environmental siege.* Karkheh Basin Brochure. Challenge Program on Water and Food (CPWF). http://www.waterandfood.org/

CPWF. 2005. *Basin focal project: Karkheh proposal.* Challenge Program on Water and Food (CPWF). http://www.waterandfood.org/

Chang, M.; Lee, R. 1974. Objective Double-Mass Analysis. Water Resources Research 10 (6): 1123-1126.

Chaplot, V. 2005. Impact of DEM mesh size and soil map scale on SWAT runoff, sediment, and NO_3–N loads predictions. *Journal of Hydrology* 312: 207-222.

Chaplot, V.; Saleh, A.; Jaynes, D.B. 2005. Effect of the accuracy of spatial rainfall information on the modeling of water, sediment, and NO_3–N loads at the watershed level. *Journal of Hydrology* 312: 223-234.

Cho, H.; Olivera, F. 2009. Effect of the spatial variability of land use, soil type, and precipitation on streamflows on small watersheds. *Journal of the American Water Resources Association* 45(3): 673-686.

Christensen, J. H.; Hewitson, B.; Busuioc, A.; Chen, A.; Gao, X.; Held, I.; Jones, R.; Kolli, R.K.; Kwon, W.-T.; Laprise, R.; Magaña Rueda, V.; Mearns, L.; Menéndez, C.G.; Räisänen, J.; Rinke, A.; Sarr, A.; Whetton, P. 2007. Regional climate projections. In: *Climate Change 2007: The Physical Science Basis.* Contribution of working group I to the fourth assessment report of the Intergovernmental Panel on Climate Change, pp. 847-940, ed. Solomon, S.; Qin, D.; Manning, M.; Chen, Z.; Marquis, M.; Averyt, K.B.; Tingnor, M.; Miller, H.L. Cambridge, UK: Cambridge University Press.

Cosgrove, W.J.; Rijsberman, F.R. 2000a. Challenge for the 21st century: Making water everybody's business. *Sustainable Development International* 2: 149-156.

Cosgrove, W.J.; Rijsberman, F.R. 2000b. *World water vision: making water everybody's business.* London, UK: Earthscan Publications.

Cudennec, C.; Leduc, C.; Koutsoyiannis, D. 2007. Dryland hydrology in Mediterranean regions-A review. *Hydrological Sciences Journal* 52 (6): 1077-1087.

Cullen, H.M.; Kaplan, A.; Arkin, P.A.; Demenocal, P.B. 2002. Impact of the North Atlantic oscillation on the Middle Eastern climate and streamflow. *Climatic Change* 55: 315-338.

Daly, C.; Gibson, W.P.; Taylor, G H.; Johnson, G.L.; Pasteris, P. 2002. A knowledge-based approach to the statistical mapping of climate. *Climate Research* 22: 99-113.

De Fries, R.; Asner, G.P.; Houghton, R., eds. 2004. Ecosystems and land use change. *Geophys. Monogr. Ser.* 153: 308, Washington, DC: American Geophysical Union.

Domroes, M.; Kaviani, M.; Schaefer, D. 1998. An analysis of regional and intra-annual precipitation variability over Iran using multivariate statistical methods. *Theoretical and Applied Climatology* 61: 151-159.

DIALOGUE. 2002. *DIALOGUE on water, food and environment: Proposal.* Dialogue Consortium: FAO, GWP, ICID, IFAP, IWMI, IUCN, UNEP, WHO,WWC and WWF. http://www.cgiar.org/iwmi/dialogue/dialogue.htm.

Dinpashoh, Y.; Fakheri-Fard, A.; Moghaddam, M.; Jahanbakhsh, S.; Mirnia, M. 2004. Selection of variables for the purpose of regionalization of Iran's precipitation climate using multivariate methods. *Journal of Hydrology* 297: 109-123.

Dixon, B.; Earls, J. 2009. Resample or not? Effects of resolution of DEMs in watershed modeling. *Hydrological Processes* 23: 1714-1724.

Döll, P. 2002. Impact of climate change and variability on irrigation requirements: A global perspective. *Climatic Change* 54: 269-293.

Douglas, E.M.; Vogel, R.M.; Kroll, C.N. 2000. Trends in floods and low flows in the United States: Impact of spatial correlation. *Journal of Hydrology* 240: 90-105.

Evans, J.P.; Smith, R.B.; Oglesby, R.J. 2004. Middle East climate simulation and dominant precipitation processes. *International Journal of Climatology* 24: 1671-1694.

Falkenmark, M.; Lundquist, J.; Widstrand, C. 1989. Macro-scale water scarcity requires micro-scale approaches: Aspects of vulnerability in semi-arid development. *Natural Resources Forum* 13(4): 258-267.

Falkenmark, M.; Rockström, J. 2006. The new blue and green water paradigm: Breaking new ground for water resources planning and management. Editorial. *Journal of Water Resources Planning and Management* 132(3): 129-132.

Falkenmark, M.; Molden, D. 2008. Wake up to realities of river basin closure. *Water Resources Development* 24(2): 201-15.

Faramarzi, M.; Abbaspour, K.C.; Schulin, R.; Yang, H. 2009. Modelling blue and green water resources availability in Iran. *Hydrological Processes* 23: 486-501.

Faramarzi, M.; Yang, H.; Schulin, R.; Abbaspour, K.C. 2010. Modeling wheat yield and crop water productivity in Iran: Implications of agricultural water management for wheat production. *Agricultural Water Management* 97: 1861-1875.

Faurès, J.M.; Goodrich, D.C.; Woolhiser, D.A.; Sorooshian, S. 1995. Impact of small-scale spatial rainfall variability on runoff modelling. *Journal of Hydrology* 173: 309–326.

Ficklin, D.L.; Luo, Y.; Luedeling, E.; Zhang, M. 2009. Climate change sensitivity assessment of a highly agricultural watershed using SWAT. *Journal of Hydrology* 374: 16-29.

FAO (Food and Agriculture Organization of the United Nations. 1995. *The Digital Soil Map of the World and Derived Soil Properties. CD-ROM, Version 3.5.* Rome, Italy: FAO. http://www.fao.org/

FAO. 1997. *Iran water resources.* Rome, Italy: FAO. http://www.fao.org/ag/agl/aglw/aquastat/countries/iran/index.stm

FAO. 2006. *AQUASTAT-FAO's information system on water and agriculture.* Rome, Italy: FAO. http://www.fao.org/

FAO. 2009. *AQUASTAT-FAO's information system on water and agriculture.* Rome, Italy: FAO. http://www.fao.org/

Foltze, R.C. 2002. Iran's water crisis: Cultural, political and ethical dimensions. *Journal of Agricultural and Environmental Ethics* 15: 357-380.

Fontaine, T.A.; Cruickshank, T.S.; Arnold, J.G.; Hotchkiss, R.H. 2002. Development of a snowfall-snowmelt routine for mountainous terrain for the Soil Water Assessment Tool (SWAT). *Journal of Hydrology* 262: 209-223.

Gallart, F.; Latron, J.; llorens, P.; Beven, K. 2006. Using internal catchment information to reduce the uncertainty of discharge and baseflow predictions, *Advances in Water Resources,* doi:10.1016/j.advwatres.2006.06.005.

Gallopin, G.C.; Rijsberman, F.R. 2000. Three global water scenarios. *International Journal of Water* 1(1): 16-40.

Ganji, A.; Khalili, D.; Karamouz, M. 2007. Development of stochastic dynamic Nash game model for reservoir operation. I. The symmetric stochastic model with perfect information. Advances in Water Resources 30: 528-542.

Gassman, P.W.; Reyes, M.R.; Green, C.H.; Arnold, J.G. 2007. The soil water assessment tool: Historical development, applications, and future research directions. *Transactions of the ASABE* 50(4): 1211-1250.

George, S.S. 2007. Streamflow in the Winnipeg River Basin, Canada: Trends, extremes and climate linkages. *Journal of Hydrology* 332: 396-411.

Ghafouri, M.; Turkleboom, F.; Sarreshtehdhari, A. 2007. Integrated catchments management – Lessons learned and Iranian vision. In: *Extended abstracts. International workshop on improving water productivity and livelihood resilience in Karkheh River Basin,* Ghafouri, M. (ed.) Tehran, Iran: Soil Conservation and Watershed Management Research Institutes (SCWNRI), September 10–11, 2007.

Githui, F.; Mutua, F.; Bauwens, W. 2009a. Estimating the impacts of land-cover change on runoff using the soil and water assessment tool (SWAT): case study of Nzoia catchment, Kenya. *Hydrological Sciences Journal* 54(5): 899-908.

Githui, F.; Gitau, W.; Mutua, F.; Bauwens, W. 2009b. Climate change impact on SWAT simulated streamflow in Western Kenya. *International Journal of Climatology* 29(12): 1623-1834. DOI 10.1002/joc.1828.

Gleick, P. 2003. Global freshwater resources: Soft path solutions for the 21st century. *Science* 302: 1524-1528.

Goswami, M.; O'Connor, K.M.; Bhattarai, K.P. 2007. Development of regionalization procedures using a multi-model approach for flow simulation in an ungauged catchment. *Journal of Hydrology* 333: 517-531.

Götzinger, J.; Bárdossy, A. 2007. Comparison of four regionalization methods for a distributed hydrologic model. *Journal of Hydrology* 333, 374-384.

Green, G.P.; Hamilton, J.R. 2000. Water allocation, transfers and conservation: links between policy and hydrology. *Water Resources Development* 16(2): 197-208.

Guo, H.; Hu, Q.; Jian, T. 2008. Annual and seasonal streamflow responses to climate and land-cover changes in the Poyang Lake Basin, China. *Journal of Hydrology* 355: 106-122.

Gupta, A.D. 2008. Implications of environmental flows in river basin management. *Physics and Chemistry of the Earth* 33 (5): 298-303.

Gupta, H.V.; Kling, H.; Yilmaz, K.K.; Martinez, G.F. 2009. Decomposition of the mean squared error and NSE performance criteria: Implications for improving hydrological modelling. *Journal of Hydrology* 377: 80-91.

Hargreaves, G.L.; Hargreaves, G.H.; Riley, J.P. 1985. Agricultural benefits for Senegal River Basin. *Journal of Irrigation and Drainage Engineering* 111: 113-124.

Heydari, N. 2006. Water and irrigation management in the water-stressed Zayandeh-Rud and Karkheh River Basins, Islamic Republic of Iran. In proceedings of the International Forum on water resources management and irrigation modernization in Shanxi Province, China, 22-24 November, 2006, Taiyuan, China.

Hughes, D.A.; Smakhtin, V.Y. 1996. Daily flow time series patching or extension: A spatial interpolation approach based on FDCs. *Hydrological Sciences Journal* 41(6): 851–71.

Ibrahim, A.B.; Cordery, I. 1995. Estimation of recharge and runoff volumes from ungauged catchments in eastern Australia. *Hydrological Sciences Journal* 40: 499-515.

IWMI. 2005. *Annual Report 2004/2005*. Colombo, Sri Lanka, International Water Management Institute (IWMI).

JAMAB. 1999. *Comprehensive Assessment of National Water Resources: Karkheh River Basin*. JAMAB Consulting Engineers in association with Ministry of Energy, Iran. (In Persian).

JAMAB. 2006a. *A report on the development of ground waters and utilization thereof in Karkheh Drainage Basin: An analysis of specifications.* JAMAB Consulting Engineers in association with Ministry of Energy, Iran.

JAMAB 2006b. *Water balance report of Karkheh River Basin area: Preliminary analysis.* JAMAB Consulting Engineers in association with Ministry of Energy, Iran.

Jayakrishnan, R.; Srinivasan, R.; Santhi, C.; Arnold, J.G. 2005. Advances in the application of the SWAT model for water resources management. *Hydrological Processes* 19: 749-762.

Jha, M.; Gassman, P.W.; Secchi, S.; Gu, R.; Arnold, J.G. 2004. Effect of watershed subdivision on SWAT flow, sediment, and nutrient predictions. *Journal of the American Water Resources Association* 40(3): 811-825.

Jones, C.; Sultan, M.; Yan, E.; Milewski, A.; Hussein, M.; Al-Dousari, A.; Al-kaisy, S.; Becker, R. 2008. Hydrologic impacts of engineering projects on the Tigris-Euphrates systems and its marshlands. *Journal of Hydrology* 353: 59-75.

Karamouz, M.; Zahraie, B.; Araghi-Nejhad, Sh.; Shahsavari, M.; Torabi, S. 2001. An integrated approach to water resources development of the Tehran region in Iran. *Journal of the American Water Resources Association* 37 (5): 1301-1311.

Karamouz, M.; Zaharie, B.; Khodatalab, N. 2003. Reservoir optimization: A non-structural solution for control of seepage from Lar Reservoir in Iran. *Water International* 28(1): 19-26.

Karamouz, M.; Akhbari, M.; Moridi, A.; Kerachian, R. 2006. A system dynamics-based conflict resolution model for river water quality management. *Iranian Journal of Environment and Health Science and Engineering* 3 (3): 147-160.

Karamouz, M.; Moridi, A.; Fayyazi, H.M. 2008. Dealing with conflict over water quality and quantity allocation: A case study. *Scientia Iranica* 15(1): 34-49.

Karamouz, M.; Akhbari, M.; Moridi, A. 2011. Resolving disputes over reservoir-river operations. Journal of Irrigation and Drainage Engineering 137(5): 327-339.

Keller, A.; Keller, J.; Seckler, D. 1996. *Integrated water resource systems: Theory and policy implications.* Research Report 3. Colombo, Sri Lanka: International Water Management Institute.

Keller, J.; Keller, A.; Davids, G. 1998. River basin development phases and implications of closure. *Journal of Applied Irrigation Science* 33(2): 145-164.

Kerachian, R.; Karmouz, M. 2007. A stochastic conflict resolution model for water quality management in reservoir-river systems. *Advances in Water Resources* 30: 866-882.

Keshavarz, A.; Asadollahi, A.; Pazira, E. 2007. Assessment of agricultural water policies, laws, and regulations influencing water productivity across KRB. In: *Extended abstracts. International workshop on improving water productivity and livelihood resilience in Karkheh River Basin,* Ghafouri, M.

(ed.) Tehran, Iran: Soil Conservation and Watershed Management Research Institutes (SCWNRI), September 10–11, 2007.

Kokkonen, T.S.; Jakeman, A.J.; Young, P.C.; Koivusalo, H.J. 2003. Predicting daily flows in ungauged catchments: Model regionalization from catchment descriptors at the Coweeta Hydrologic Laboratory, North Carolina. *Hydrological Processes* 17: 2219–2238.

Krause, P.; Boyle, D.P.; Base, F. 2005. Comparison of different efficiency criteria for hydrological model assessment. *Advances in Geosciences* 5: 89-97.

Lacombe, G.; Cappelaere, B.; Leduc, C. 2008. Hydrological impact of water and soil conservation works in the Merguellil catchment of Central Tunisia. *Journal of Hydrology* 359: 210-224.

Larocque, M.; Fortin, V.; Pharand, M.C.; Rivard, C. 2010. Groundwater contribution to river flows using hydrograph separation, hydrological models in a southern Quebec aquifer. *Hydrology and Earth System Sciences Discussion* 7: 7809-7838.

Lettenmaier, D.P.; Wood, E.F.; Wallis, J.R. 1994. Hydro-climatological trends in the Continental United States, 1948-88. *Journal of Climate* 7: 586-607.

Lidén, R.; Harlin, J. 2000. Analysis of conceptual rainfall-runoff modelling performance in different climates. *Journal of Hydrology* 238: 231-247.

Linsley, R.K.; Kohler, M.A.; Paulhus, J.L.H. 1949. *Applied Hydrology.* McGraw-Hill, New York.

Linsley, R.K.; Kohler, M.A.; Paulhus, J.L.H. 1982. *Hydrology for engineers (3rd ed).* McGraw-Hill series in Water Resources and Environmental Engineering, McGraw-Hill Inc., New York.

Lim, J.K.; Engel, B.A.; Tang, Z.; Kim, J.C. 2005. Automated web GIS based hydrograph analysis tool, WHAT. *Journal of the American Water Resources Association* 41(6): 1407-1416.

Loucks, D.P. 2006. Modeling and managing the interactions between hydrology, ecology and economics. *Journal of Hydrology* 328: 408- 416.

Love, D.; Uhlenbrook, S.; Corzo-Perez, G.; Twomlow, S. 2010. Rainfall-interception-evaporation-runoff relationships in a semi-arid meso-catchment, northern Limpopo Basin, Zimbabwe. *Hydrological Sciences Journal* 55 (5): 687-703.

Magette, W.L.; Shanholtz, V.O.; Carr, J.C. 1976. Estimating selected parameters for the Kentucky watershed model from watershed characteristics. *Water Resources Research* 1: 472-476

Makurira, H.; Savenije, H.H.G.; Uhlenbrook, S. 2010. Modeling field scale water partitioning using on-site observations in sub-Saharan rainfed agriculture. *Hydrology and Earth System Sciences* 14, 627-638.

Mann, M.E. 2002. Large-scale climate variability and connections with the Middle East in the past centuries. *Climatic Change* 55: 287-314.

Manouchehri, G.R.; Mahmoodian, S.A. 2002. Environmental Impacts of dams constructed in Iran. *International Journal of Water Resources Development* 18(1): 179-182.

Marjanizadeh, S. 2008. Developing a "best case scenario"for Karkheh River Basin management (2025 horizon); a case study from Karkheh River Basin, Iran. PhD dissertation, Department of Water, Atmosphere and Environment, University of Natural Resources and Applied Life Sciences, Vienna, Austria.

Marjanizadeh, S.; Qureshi, A.S.; Turral, H.; Talebzadeh, P. 2009. *From Mesopotamia to the third millennium: The historical trajectory of water development and use in the Karkheh River Basin, Iran.* IWMI Working Paper 135. Colombo, Sri Lanka: International Water Management Institute. . doi:10.3910/2010.206.

Marjanizadeh, S.; de Fraiture, C.; Loiskandl, W. 2010. Food and water scenarios for the Karkheh River Basin, Iran. *Water International* 35 (4): 409-424.

Masih, I.; Ahmad, M.D.; Turral, H.; Uhlenbrook, S.; Karimi, P. 2009. Analysing streamflow variability and water allocation for sustainable management of water resources in the semi-arid Karkheh River Basin, Iran. *Physics and Chemistry of the Earth* 34 (4-5): 329–340.

Masih, I.; Uhlenbrook, S.; Maskey, S.; Ahmad, M.D. 2010. Regionalization of a conceptual rainfall-runoff model based on similarity of the flow duration curve: A case study from the semi-arid Karkheh basin, Iran, *Journal of Hydrology* 391: 188-201. DOI:10.1016/j.jhydrol.2010.07.018

Masih, I.; Maskey, S.; Uhlenbrook, S.; Smakhtin, V. 2011a. Assessing the impact of areal precipitation input on streamflow simulations using the SWAT model. *Journal of the American Water Resources Association* .(JAWRA) 47(1):179-195. DOI: 10.1111/j.1752-1688.2010.00502.x.

Masih, I.; Uhlenbrook, S.; Maskey, S.; Smakhtin, V. 2011b. Streamflow trends and climate linkages in the Zagros Mountain, Iran. *Climatic Change* 104: 317-338. DOI 10.1007/s10584-009-9793-x.

Masih, I.; Maskey, S.; Uhlenbrook, S.; Smakhtin, V. 2011c. Quantifying scale-dependent impacts of upgrading rain-fed agriculture in a semi-arid basin. *Agricultural Water Management* (in review).

Maskey, S.; Guinot, V.; Price, R.K. 2004. Treatment of precipitation uncertainty in rainfall-runoff modelling: A fuzzy set approach. *Advances in Water Resources* 27: 889-898.

Maskey, S. 2007. *HyKit - A tool for grid based interpolation of hydrological variables, user's guide.* Delft, the Netherlands: UNESCO-IHE Institute for Water Education.

Mathews, E. 1983. Global vegetation and land use: new high-resolution data bases for climate studies, *Journal of Climate and Applied Meteorology* 22: 474–487.

McCartney, M.; Smakhtin, V. 2010. *Water storage in an era of climate change: Addressing the challenge of increasing rainfall variability.* Blue Paper. International Water Management Institute, Colombo, Sri Lanka.

McCuen, R.H. 2003. *Modelling hydrologic change: Statistical methods.* Boca Raton, Florida: Lewis Publishers.

McIntyre, N.; Lee, H.; Wheater, H.S.; Young, A.; Wagener, T. 2005. Ensemble Predictions of runoff in ungauged catchments. *Water Resources Research,* 41: W12434, doi:10.1029/2005WR004289.

Merz, B.; Blöschl, G. 2004. Regionalization of catchment model parameters. *Journal of Hydrology* 287: 95-123.

Mirqasemi, S.A.; Pauw, E. De. 2007. Land use change detection in the Karkheh Basin, Iran by using multi-temporal satellite images. In: *Extended abstracts. International Workshop on Improving Water Productivity and Livelihood Resilience in Karkheh River Basin,* ed. Ghafouri, M. Soil Conservation and Watershed Management Research Institutes (SCWNRI), Tehran, Iran. September 10-11, 2007.

Mitchell, T.D.; Carter, T.R.; Jones, P.D.; Hulme, M.; New, M. 2004. *A comprehensive set of high-resolution grids of monthly climate for Europe and the globe: The observed record (1901–2000) and 16 scenarios (2001– 2100).* Working Paper 55, Tyndall Centre.

Modallaldoust, S.; Bayat, F.; Soltani, B.; Soleimani, K. 2008. Applying digital elevation model to interpolate precipitation. *Journal of Applied Sciences* 8(8): 1471-1478.

Modarres, R.; da Silva, V. de P.R. 2007. Rainfall trends in arid and semi-arid regions of Iran. *Journal of Arid Environments* 70: 344-355.

Molden, D. 1997. *Accounting for water use and productivity.* SWIM Paper 1. Colombo, Sri Lanka: International Water Management Institute.

Molden, D.; Sakthivadivel, R. 1999. Water accounting to assess use and productivity of water. *Water Resources Development* 15: 55-71.

Molle, F. 2003. *Development trajectories of river basins: A conceptual framework.* Research Report 72. Colombo, Sri Lanka: International Water Management Institute.

Molle, F.; Mamanpoush, A.; Miranzadeh, M. 2004. *Robbing Yadullah's water to irrigate Saeid's garden: Hydrology and water rights in a village of central Iran.* Research Report 80. Colombo, Sri Lanka: International Water Management Institute.

Molle, F.; Turral, H. 2004. Demand management in a basin perspective: Is the potential for water saving overrated? Paper presented at the International Conference on W*ater Demand Management,* June 2004, Dead Sea, Jordan.

Molle, F. 2006. *Planning and managing water resources at the river-basin scale: Emergence and evolution of a concept.* IWMI Comprehensive Assessment Research Report 16. Colombo, Sri Lanka: International Water Management Institute.

Mousavi, S.J.; Moghaddam, K.S.; Seifi, A. 2004. Application of an interior-point algorithm for optimization of a large-scale reservoir system. Water Resources Management 18: 519-540.

Mul, M.; Savenije, H.H.G.; Uhlenbrook, S. 2009. Spatial rainfall variability and runoff response during an extreme event in a semi-arid, meso-scale catchment in the South Pare mountains, Tanzania. *Hydrology and Earth System Sciences* 13: 1659–1670.

Mul, M.L.; Mutiibwa, R.K.; Uhlenbrook, S.; Savenije, H.H.G. 2008. Hydrograph separation using hydrochemical tracers in the Makanya catchment, Tanzania. *Physics and Chemistry of the Earth* 33 (1): 151-156.

Muthuwatta, L.P.; Ahmad, M.D.; Bos, M.G.; Rientjes, T.H.M. 2010. assessment of water availability and consumption in the Karkheh River Basin, Iran— Using remote sensing and geo-statistics. *Water Resources Management* 24: 459-484, DOI 10.1007/s11269-009-9455-9.

Nash, J.E.; Sutcliffe, J.V. 1970. River flow forecasting through conceptual models: I. A discussion of principles. *Journal of Hydrology* 10: 282-290.

Nash, L.L.; Gleick, P.H. 1991. Sensitivity of streamflow in the Colorado Basin to climatic changes. *Journal of Hydrology* 125: 221-241.

Nathan, R.J.; McMahon, T.A. 1990a. Evaluation of automated techniques for base flow and recession analyses. *Water Resources Research* 26:1465–1473.

Nathan, R.J.; McMahon, T.A. 1990b. Identification of homogeneous regions for the purpose of regionalization. *Journal of Hydrology* 121: 217-238.

Nazemosadat, M.J.; Cordery, I. 2000. On the relationships between ENSO and autumn rainfall in Iran. *International Journal of Climatology* 20: 47-61.

Neitsch, S.L.; Arnold, J.G.; Kiniry, J.R.; Williams, J.R. 2005. *Soil and water assessment tool theoretical documentation: Version 2005*. Texas, USA: USDA, Soil and Water Research Laboratory/Blackland Research Center.

New, M.; Hulme, M.; Jones, P.D. 2000. Representing twentieth century space-time climate variability part 2. Development of 1901–1996 monthly grids of terrestrial surface climate. *Journal of Climate* 13: 2217–2238.

Newman, B.D.; Wilcox, B.P.; Archer, S.R.; Breshears, D.D.; Dahm, C.N.; Duffy, C.J.; McDowell, N.G.; Phillips, F.M.; Scanlon, B.R.; Vivoni, E.R. 2006. Ecohydrology of water-limited environments: A scientific vision. *Water Resources Research* 42, W06302, doi:10.1029/2005WR004141.

Niadas, I.A.; Mentzelopoulos, P.G. 2008. Probabilistic flow duration curves for small hydro plant design and performance evaluation. *Water Resources Management* 22: 509-523

Oudin, L.; Andreassian, V.; Perrin, C.; Michel, C.; Le Moine, N. 2008. Spatial proximity, physical similarity, regression and ungauged catchments: A comparison of regionalization approaches based on 913 French catchments. *Water Resources Research* 44: W03413, doi: 10.1029/2007WR006240.

Oudin, L.; Perrin, C.; Mathevet, T.; Andreassian, V.; Michel, C. 2006. Impact of biased and randomly corrupted inputs on the efficiency and the parameters of watershed models. *Journal of Hydrology* 320 (1-2): 62-83.

Oweis, T.; Hachum, A. 2009. Optimizing supplemental irrigation: Tradeoffs between profitability and sustainability. *Agricultural Water Management* 96: 511-516.

Palmer, M.A.; Bernhardt, E.S. 2006. Hydroecology and river restoration: Ripe for research and synthesis. *Water Resources Research* 42: W03S07, doi:10.1029/2005WR004354.

Parajka, J.; Merz, R.; Blöschl, G. 2005. A comparison of regionalisation methods for catchment model parameters. *Hydrology and Earth System Sciences Discussion* 2: 509-542.

Parsons, A.J.; Abrahams, A. D. 1994. Geomorphology of desert environments. In: *Geomorphology of Desert Environments,* ed. Abrahams, A.D.; Parsons, A.J. Boca Raton, Fla.: CRC Press, pp. 1-12.

Poff, N.L.; Allan, J.D.; Bain, M.B.; Karr, J.R.; Prestegaard, K.L.; Richter, B.D.; Sparks, R.E.; Stromberg, J.C. 1997. The natural flow regime. *Bioscience* 47(11): 769- 84.

Postel, S. 2000. Entering an era of water scarcity: The challenges ahead. *Ecological Applications* 10 (4): 941-948.

Postel, S. 2003. Securing water for people, crops, and ecosystems: new mindset and new priorities. *Natural Resource Forum* 47: 89-98.

Postel, S. 2005. *Liquid assets: The critical need to safeguard freshwater ecosystems.* Washington, DC: Worldwatch Institute.

Postel, S.; Richter, B. 2003. *Rivers for life: Managing water for people and nature.* Washington, DC: Island Press.

Quenouille, M.H. 1956. Notes on bias in estimation. *Biometrika* 43: 353-360.

Qureshi, A.S.; Ahmad, M.D.; Gichuki, F.; Clemet, F.; Masih, I. 2005. Diagnostic tour of the Karkheh River Basin: General observations and suggestions for basin focal project priorities. Interim Project Report, Basin Focal Project. International Water Management Institute, Karaj, Iran. (Duplicated).

Richter, B.D.; Baumgartner, J.V.; Wigington, R.; Braun, D.P. 1997. How much water does a river need? *Freshwater Biology* 37: 231-249.

Rijsberman, F.R. 2006. Water scarcity: Fact or fiction? *Agricultural Water Management* 80: 5-22.

Rijsberman, F.R.; Molden, D.J. 2001. Balancing water uses: Water for food and water for nature. Thematic background papers. *International Conference on Freshwater.* Bonn, 3-7 December, pp. 43-56.

Ringeltaube, J. 2002. The European Water Framework Directive-An example for water management in national and international river basins. In: *FRIEND 2002- Regional hydrology: Bridging the Gap between research and practices,* Proceedings of the fourth international FRIEND conference held at Cape Town, South Africa, March 2002. IAHS publ. no. 274. 2002. pp. 19-25.

Rockström, J.; Karlberg, L.; Wani, S.P.; Barron, J.; Hatibu, N,; Oweis, T.; Bruggeman, A.; Farahani, J.; Qiang, Z. 2010. Managing water in rain-fed agriculture-the need for a paradigm shift. *Agricultural Water Management* 97: 543-550.

Saghafian, B.; Davtalab, R. 2007. Mapping snow characteristics based on snow observation probability. *International Journal of Climatology* 27: 1277-1286.

Santhi, C.; Kannan, N.; Arnold, J.G.; Di Luzio, M. 2008. Spatial calibration and temporal validation of flow for regional scale hydrologic modeling. *Journal of the American Water Resources Association* 44(4):829-846.

Savenije, H.H.G.; van der Zaag, P. 2008. Integrated water resources management: Concepts and issues. *Physics and Chemistry of the Earth* 33: 290-297.

Seckler, D. 1996. *The new era of water resources management: From "dry" to "wet" water savings*. Research Report 1. Colombo, Sri Lanka: International Irrigation Management Institute.

Seckler, D.; Amarasinghe, U.; Molden, D.J.; de Silva, R.; Barker, R. 1998. *World water demand and supply, 1990 to 2025: scenarios and issues*. IWMI Research Report 19. Colombo, Sri Lanka: International Water Management Institute.

Seibert, J. 1999. Regionalization of parameters for a conceptual rainfall-runoff model. *Agriculture and Forest Meteorology* 98-99: 279-293.

Seibert, J. 2000. Multi-criteria calibration of a conceptual rainfall-runoff model using a genetic algorithm. *Hydrology and Earth System Sciences* 4 (2): 215-224.

Seibert, J. 2002. *HBV light version 2, user's manual*. Oregon, USA: Oregon State University Department of Forest Engineering Corvallis. Uppsala: Uppsala University, Dept. of Earth Science, Hydrology.

Servat, E.; Dezetter, A. 1993. Rainfall-runoff modelling and water resources assessment in northwestern Ivory Coast. Tentative extension to ungauged catchments. *Journal of Hydrology* 148: 231-248.

Shiklomanov, I.A. 1999. *World water resources and water use: present assessment and outlook for 2025*. St. Petersburg, Russia: State Hydrological Institute.

Shiklomanov, I.A.; Rodda, J.C. 2003. *World water resources at the beginning of the 21st century*. Cambridge, UK: Cambridge University Press.

Schultze, G.A. 2001. Integrated water resources management: The requirements of the European Union, the problem of environmental impact assessment, and implementation of the sustainable development principle. In: *Integrated water resources management*, Proceedings of a symposium held at Davis, California, April 2000. IAHS Publ. no. 272. 2001, pp. 3-11.

Sivapalan, M.; Takeuchi, K.; Franks, S.W.; Gupta, V.K.; Karambiri, H.; Lakshmi, V.; Liang, X.; McDonnell, J.J.; Mendiondo, E.M.; O'Connell, P.E.; Oki, T.; Pomeroy, J.W.; Schertzer, D.; Uhlenbrook, S.; Zehe, E. 2003. IAHS decade on predictions in ungauged basins (PUB), 2003-2012: Shaping an exciting future for the hydrological sciences. *Hydrological Sciences Journal* 48(6): 857-880.

Smakhtin, V.U. 2001a. Low flow hydrology: A review. *Journal of Hydrology* 240: 147-186

Smakhtin, V.U. 2001b. Estimating continuous monthly baseflow time series and their possible applications in the context of the ecological reserve. *Water SA* 27 (2): 213-218.

Smakhtin, V.U.; Revenga, C.; Döll, P. 2004. *Taking into account environmental water requirements in global-scale water resources assessments*. Comprehensive Assessment of Water Management in Agriculture Research Report 2. Colombo, Sri Lanka: International Water Management Institute.

Snellen, W.B.; Schrevel, A. 2004. *IWRM for sustainable use of water: 50 years of international experience with the concept of integrated water resources management.* Background document to the FAO/Netherlands Conference on Water for Food and Ecosystems. Wageningen, Alterra, Alterra-report 1143.

Soil Conservation Service Engineering Division. 1986. *Urban hydrology for small watersheds.* Technical Release 55. Washington, DC: US Department of Agriculture.

Srinivasan, R.; Ramanarayanan, T.S.; Arnold, J.G.; Bednarz, S.T. 1998. Large area hydrologic modeling and assessment part II: Model application. *Journal of the American Water Resources Association* 34(1): 91-101.

Starks, P.J.; Moriasi, D.N. 2009. Spatial resolution effect of precipitation data on SWAT calibration and performance: Implications for CEAP. *Transactions of the ASABE* 52(4): 1171-1180.

Sullivan, C.A.; Meigh, J.R.; Simon, S.; Lawrence, P.; Calow, R.; McKenzie, A.; Acreman, M.C.; Moore, R.V. 2000. *The development of a water poverty index.* Report to DFID. Wallingford, UK: Centre for Ecology and Hydrology.

Sutcliffe, J.V. 2004. *Hydrology: A question of balance.* IAHS special publication 7. Wallingford, UK: IAHS Press.

Sutcliffe, J.V.; Carpenter, T.G. 1968. The assessment of runoff from a mountainous and semi-arid area in western Iran. In: *Hydrological Aspects of the Utilization of Water.* IAHS General Assembly of Bern, IAHS Publ. 76. Wallingford, UK: IAHS Press, pp. 383-394.

Tennant, D.L. 1976. Instream flow regimens for fish, wildlife, recreation and related environmental resources. *Fisheries* 1: 6-10.

Tetzlaff, D.; Uhlenbrook, S. 2005. Effects of spatial variability of precipitation for process-oriented hydrological modelling: Results from two nested catchments. *Hydrology and Earth System Sciences* 9: 29-41.

Tharme, R. 2003. A global perspective on environmental flow assessment: Emerging trends in the development and application of environmental flow methodologies for rivers. *River Research and Applications* 19: 397-441.

Tizro, A.T.; Voudouris, K.S.; Eini, M. 2007. Groundwater balance, safe yield and recharge feasibility in a semi-arid environment: A case study from western part of Iran. *Journal of Applied Sciences* 7(20): 2967-2976.

Tobin, K.J.; Bennett, M.E. 2009. Using SWAT to model streamflow in two river basins with ground and satellite precipitation data. *Journal of the American Water Resources Association* 45(1): 253-271.

Tripathi, M.P.; Raghuwanshi, N.S.; Rao, G.P. 2006. Effect of watershed subdivision on simulation of water balance components. *Hydrological Processes* 20: 1137-1156.

Tu, M. 2006. Assessment of the effects of climate variability and land use change on the hydrology of the Meuse River Basin. PhD thesis. UNESCO-IHE Institute for Water Education, Delft, the Netherlands and Vrije Universiteit, Amsterdam, the Netherlands. London, UK: Taylor & Francis Group.

Uhlenbrook, S. 2006. Catchment hydrology with satellites, models and rubber boots. Inaugural Address of Stefan Uhlenbrook, Professor of Hydrology, Head of the Hydrology and Water Resources Core at the UNESCO-IHE Institute for Water Education in Delft, The Netherlands

Uhlenbrook, S.; Leibundgut, C. 2002. Process-oriented catchment modelling and multiple-response validation. *Hydrological Processes* 16: 423-440.

Uhlenbrook, S.; Frey, M.; Leibundgut, C.; Maloszewski, P. 2002. Hydrograph separations in a meso scale mountainous basin at event and seasonal time scales. *Water Resources Research* 38(6): 1-14.

Uhlenbrook, S.; Seibert, J.; Leibundgut, C.; Rodhe, A. 1999. Prediction uncertainty of conceptual rainfall-runoff models caused by problems to identify model parameters and structure. *Hydrological Sciences Journal* 44 (5): 779-797.

Uhlenbrook, S.; Franks, S.; Heal, K.; Hubbard, S.; Karambiri, H.; Oki, T.; Valeo, C. 2006. Key messages, recommendations and concluding remarks. Chapter 8. *Hydrology 2020: An integrating science to meet world water challenges,* ed. Oki, T.; Valeo, C.; Heal, K. IAHS Publication 300. Wallingford, UK: IAHS Press.

UNEP. 2001. Partow, H. *The Mesopotamian Marshlands: Demise of an Ecosystem.* Early warning and assessment technical report, UNEP/DEWA/TR.01-3 Rev.1, Division of Early Warning and Assessment, United Nations Environment Program, Nairobi, Kenya.

Vakili, A.; Mousavi, S.F.; Karamooz, M., eds. 1995. An overview of water resources development in Iran. *Proceedings of Regional Conference on Water Resources Management.* Conference Secretariat; Isfahan University of Technology, Isfahan, Iran.

van der Zaag, P. 2005. Integrated water resources management: Relevant concept or irrelevant buzzword? A capacity building and research agenda for Southern Africa. *Physics and Chemistry of the Earth* 30: 867-871.

van der Zaag, P.; Gupta, J. 2008. Scale issues in the governance of water storage projects. *Water Resources Research* 44, W10417, doi:10.1029/2007WR006364.

Vandenberghe, V.; van Griensven, A.; Bauwens, W.; Vanrolleghem, P.A. 2005. Propagation of uncertainty in diffuse pollution into water quality predictions: application to the River Dender in Flanders, Belgium. *Water Science & Technology* 51(3-4): 347-354.

Vandewiele, G.L.; Xu, C.Y.; Huybrechts, W. 1991. Regionalisation of physically-based water balance models in Belgium: application to ungauged catchments. *Water Resources Management* 5: 199-208.

van Griensven, A.; Bauwens, W. 2003. Multi-objective auto-calibration of semi-distributed water quality models. *Water Resources Research* 39 (12): 1348, DOI 10.1029/2003WR002284.

van Griensven, A.; Meixner, T.; Grunwald, S.; Bishop, T.; Diluzio, M.; Srinivasan, R. 2006. A Global sensitivity analysis tool for the parameters of multi-variable catchment models. *Journal of Hydrology* 324: 10-23.

Vogel, R.M.; Fennessey, N.M. 1995. Flow duration curves. II. a review of application in water resource planning. *Water Resources Bulletin* 31 (6): 1029-1039.

von Storch; Navarra, A., eds. 1995. *Analysis of climate variability*. New York: Springer.

Vörösmarty, C.J.; Green, P.; Salisbury, J.; Lammers, R.B. 2000. Global water resources: Vulnerability from climate change and population growth. *Science* 289:284-288.

Wagener, T.; Wheater, H.S.; Gupta, H.V. 2004. Rainfall-runoff modelling in gauged and ungauged catchments. London, UK: Imperial College Press.

Wagener, T.; Wheater, H.S. 2006. Parameter estimation and regionalization for continuous rainfall-runoff models including uncertainty. *Journal of Hydrology* 320: 132-54.

Wagener, T.; Sivapalan, M.; Troch, P.; Woods, R. 2007. Catchment classification and hydrologic similarity. *Geography Compass* 1, 10.1111/j.1749-8198.2007.00039.x

Wakindiki, I.I.C.; Ben-Hur, M. 2002. Indigenous soil and water conservation techniques: Effects on runoff, erosion, and crop yields under semi-arid conditions. *Australian Journal of Soil Research* 40: 367-379.

Wallace, J. S. 2000. Increasing agricultural water efficiency to meet future food production. *Agriculture, Ecosystems and Environment* 82: 105-119.

Wang, Q.J. 1991. The genetic algorithm and its application to calibrating conceptual rainfall-runoff models. *Water Resources Research* 27: 2467-2471.

Watson, B.M.; Srikanthan, R.; Selvalingam, S.; Ghafouri, M. 2005. Evaluation of three daily rainfall generation models for SWAT. *Transactions of the ASAE* 48(5): 1697-1711.

Weber, A.; Fohrer, N.; Möller, D. 2001. Long-term land use changes in a mesoscale watershed due to socio-economic factors — effects on landscape structures and functions. *Ecological Modelling* 140: 125-140.

Williams, J.R. 1969. Flood routing with variable travel time or variable storage coefficients. *Transactions of the ASAE* 12(1):100-103.

Welderufael, W.A.; Woyessa, Y.E. 2010. Stream flow analysis and comparison of base flow separation methods: Case study of the Modder River Basin in Central South Africa. *European Water* 31: 3-12.

Winchell, M.; Srinivasan, R.; Di. Luzio, M.; Arnold, J.G. 2008. *ARCSWAT 2.0 Interface for SWAT2005-user's guide.* Blackland Research Center, Texsas Agricultural Experiment Station and Grassland, Soil and Water Research Laboratory, USDA Agricultural Research Service Temple, Texsas.

WHO (World Health Organization). 2000. *Global water supply and sanitation assessment 2000 report.* http://www.who.int/water_sanitation_health/Globassessment/GlobalTOC.ht m. WWAP (World Water Assessment Program). (2006). *Water: A Shared Responsibility.* The United Nations World Water Development Report 2.

Xiubin, H.; Zhanbin, L.; Mingde, H.; Keli, T.; Fengli, Z. 2003. Down-scale analysis of water scarcity in response to soil-water conservation on Loess Plateau of China. *Agriculture, Ecosystems and Environment* 94: 355-361.

Yang, H.; Reichert, P.; Abbaspour, K.; Zehnder, A.J.B. 2003. A water resources threshold and its implications for food security. *Environmental Science and Technology* 37:3048-3054.

Yang, J.; Reichert, P.; Abbaspour, K.C.; Xia, J.; Yang, H. 2008. Comparing uncertainty analysis techniques for a SWAT application to the Chaohe Basin in China. *Journal of Hydrology* 358: 1-23.

Yadav, M.; Wagener, T.; Gupta, H. 2007. Regionalization of constraints on the expected watershed response behavior for improved predictions in ungauged basins. *Advances in Water Resources* 30: 1756-1774.

Yilmaz, K.K.; Gupta, H.V.; Wagener, T. 2008. A process-based diagnostic approach to model evaluation: Application to the NWS distributed hydrologic model. *Water Resources Research* 44: W09417, doi: 10.1029/2007WR006716.

Yue, S.; Wang, C.Y. 2002. Applicability of prewhitening to eliminate the influence of serial correlation on the Mann-Kendall test. *Water Resour. Res.* 38 (6), 1068, doi:10.1029/2001WR000861.

Yue, S.; Pilon, P.; Cavadias, G. 2002. Power of the Mann-Kendall and Sprearman's rho Tests for detecting monotonic trends in hydrological series. *Journal of Hydrolgy* 59: 254-271.

Zahraie, B.; Kerachian, R.; Malekmohammadi, B. 2008. Reservoir operation optimization using adaptive varying chromosome length genetic algorithm. *Water International* 33(3): 380-391.

Zahraie, B.; Hosseini, S.M. 2009. Development of reservoir operation policies considering variable agricultural water demands. Expert Systems with Applications 36: 4980-4987.

Zangvil, A.; Karas, S.; Sasson, A. 2003. Connection between eastern Mediterranean seasonal mean 500 hPa height and sea-level pressure patterns and the spatial rainfall distribution over Israel. *International Journal of Climatology* 23: 1567-1573.

Zehnder, A.J.B.; Yang, H.; Schertenleib, R. 2003. Water issues: The need for action at different levels. *Aquatic Sciences* 65: 1-20.

Zhang, X.; Harvey, K.D.; Hogg, W.D.; Yuzyk, T.R. 2001. Trends in Canadian streamflow. *Water Resources Research* 37: 987-998.

LIST OF FIGURES

APPENDIX

Appendix A. Short description of the Hargreaves method and its application in the study basin

The Hargreaves equation is commonly used for estimating reference evapotranspiration when limited amount of climatic data is available. This empirical method requires only temperature data to estimate. The Hargreaves equation is as follow (Hargreaves et al. 1985).

$$ET_o = 0.0023(T_{mean} + 17.8)(T_{max} - T_{min})^{0.5} R_a \qquad (19)$$

Where ET_o refers to reference evapotranspiration, expressed in mm/d, T_{mean}, T_{max} and T_{min} are daily mean, maximum and minimum air temperatures, expressed in °C, R_a is extraterrestrial ration, expressed here in mm/d.

The results of the Hargreaves methods were compared with the FAO Penman-Monteith method (Allen et al. 1998) using daily climatic data for the period January 1987 to December 2000 for the Kermanshah climatic station (Figure 43). The results of the both methods were found in close agreement with each other. On average, the Hargreaves method underestimated annual total ET_0 by an amount of about 5% compared to those of the FAO Penman-Monteith method. Considering these small differences, the Hargreaves method was considered appropriate to use in the study basin where limited climatic data was available.

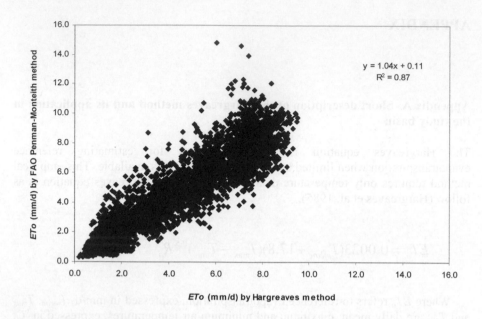

$y = 1.04x + 0.11$
$R^2 = 0.87$

Figure 43. Comparison of the estimated reference evapotranspiration (ET$_0$) by using the Hargreaves method and the FAO Penman-Monteith method at the Kermanshah climatic station in the Karkheh Basin, Iran.

ABOUT THE AUTHOR

Ilyas Masih was born in 1975 in Baddomalhi, Narowal, Pakistan. He obtained his BSc Agricultural Engineering degree in 1997 from University of Agriculture Faisalabad, Pakistan and completed his MPhil degree in Water Resources Management in 2000 from Centre of Excellence in Water Resources Engineering, University of Engineering and Technology, Lahore, Pakistan.

Ilyas has over ten years of experience as a researcher in the field of hydrology and water resources management. He has worked at IWMI from September 2001 to March 2011. He has worked at IWMI offices in Pakistan, Iran and Sri Lanka. During his professional career, he has worked on wide range of issues. Most of his undertakings involved close interaction with diversified teams of individuals representing various professional disciplines and different cultures. He has extensive experience on issues related to rainfall-runoff modeling at catchment to basin scales, analysis of long term variability and trends in the climate and streamflows, water allocation analysis and trade-offs between upstream uses and water availability for the downstream uses including environmental flow requirements, water balance and water productivity assessments in different agro-ecosystems, groundwater monitoring, evaluation and sustainable management, conjunctive use of surfacewater and groundwater resources for irrigation, secondary salinization of soils, water savings in rice-wheat cropping systems, scale considerations in up scaling water management interventions, and participatory water resources management. The use of rigorous scientific methods, collection of field and secondary data, application of analytical tools and hydrological/water management models, and synthesis of results for formulating meaningful conclusions for scientists, policy makers and other stakeholders are prominent features of his professional work. IWMI awarded him a PhD research fellowship in 2006 to undertake PhD studies as an IWMI research staff member. He was enrolled for the PhD studies in 2006 at UNESCO-IHE, Institute of Water Education, Delft, the Netherlands. His PhD research is on the issues of basin scale hydrology and water resources management in the Karkheh Basin, Iran.

Ilyas has co-authored a number of scientifically important and practically relevant papers (see selected publications below) and has presented his research at various national and international workshops and conferences.

He has recently joined UNESCO-IHE as lecturer in water resources planning. Ilyas is married to Huma and they have a daughter Sarah Ilyas.

Selected Publications

Masih, I.; Maskey, S.; Uhlenbrook, S.; Smakhtin, V. 2011. Assessing the impact of areal precipitation input on streamflow simulations using the SWAT model. *Journal of the American Water Resources Association* .(JAWRA) 47(1):179-195. DOI: 10.1111/j.1752-1688.2010.00502.x.

Masih, I.; Uhlenbrook, S.; Maskey, S.; Smakhtin, V. 2011. Streamflow trends and climate linkages in the Zagros Mountain, Iran. *Climatic Change* 104: 317-338. DOI 10.1007/s10584-009-9793-x.

Masih, I.; Uhlenbrook, S.; Maskey, S.; Ahmad, M.D. 2010. Regionalization of a conceptual rainfall-runoff model based on similarity of the flow duration curve: A case study from the semi-arid Karkheh basin, Iran, *Journal of Hydrology* 391: 188-201. DOI:10.1016/j.jhydrol.2010.07.018.

Masih, I.; Ahmad, M.D.; Turral, H.; Uhlenbrook, S.; Karimi, P. 2009. Analysing streamflow variability and water allocation for sustainable management of water resources in the semi-arid Karkheh River Basin, Iran. *Physics and Chemistry of the Earth* 34 (4-5): 329–340.

Ahmad, M.D.; Islam, Md. A.; Masih, I.; Muthuwatta, L. P.; Karimi, P.; Turral, H. 2009. Mapping basin-level water productivity using remote sensing and secondary data in the Karkheh River Basin, Iran. *Water International* 34(1):119-133.

Ahmad, M.D.; Giordano, M.; Turral, H.; Masih, I.; Masood, Z. 2007. At what scale does water saving really save water? *Journal of Soil and Water Conservation* 62(2):29A-35A.

Qureshi, A.S.; Masih, I.; Turral, H. 2006. Comparing land and water productivities of transplanted and direct dry seeded rice for Pakistani Punjab. *Journal of Applied Irrigation Science* 41(1): 47-60.

Humphreys, E.; Meisner, E.; Gupta, R.; Timsina, J.; Beecher, H.G.; Lu, T.Y.; Sing, Y.; Gill, M.A.; Masih, I.; Guo, Z.J.; Thompson, J.A. 2005. Water savings in rice-wheat systems. *Plant Production Science* 8(3): 242-258.

Ahmad, M.D.; Masih, I.; Turral, H. 2004. Diagnostic analysis of spatial and temporal variations in crop water productivity: A field scale analysis of the rice-wheat cropping system of Punjab, Pakistan. *Journal of Applied Irrigation Science* 39(1):43-63.

Qureshi, A.S. Asghar, M.N.; Ahmed, S.; Masih, I. 2004. Sustaining crop production in saline groundwater areas: A case study from Pakistani Punjab. *Australian Journal of Agricultural Research* 55:421-431.

Ahmad, M.D.; Turral, H.; Masih, I.; Giordano, M.; Masood, Z. 2007. *Water saving technologies:myths and realities revealed in Pakistan's rice-wheat systems.* Research Report 108. Colombo, Sri Lanka: International Water Management Institute.

Jehangir, W.A.; Masih, I.; Ahmed, S.; Gill, M.A.; Ahmad, M.; Mann, R.A.; Chaudhary, M.R.; Qureshi, A.S.; Turral, H. 2007. *Sustaining crop water productivity in rice-wheat systems of South Asia: A case study from Punjab Pakistan.* IWMI Working paper 115. Colombo, Sri Lanka: International Water Management Institute.

Qureshi, A.S.; Turral, H.; Masih, I. 2004. *Strategies for the management of conjunctive use of surface water and groundwater resources in semi-arid areas: A case study from Pakistan.* IWMI research report 86. Colombo, Sri Lanka: International Water Management Institute.

T - #0102 - 071024 - C180 - 244/170/10 - PB - 9780415689816 - Gloss Lamination